AF400267

Ion exchange resins and
synthetic adsorbents in food processing

Ion exchange resins and synthetic adsorbents in food processing

E.J.Zaganiaris

consultant

Second Edition

IN THE SAME SERIES::

Ion Exchange Resins in Uranium Hydrometallurgy,2nd edition,
Books on Demand, 2017

Ion Exchange Resins and Adsorbents in Chemical Processing
2nd Edition, Books on Demand, 2016

Published by:
Books on Demand GmbH
12/14 Rond Point des Champs Elysées
75008 Paris, France

Printed by:
Books on Demand GmbH, Norderstedt, Germany

© Copyright 2019: Emmanuel Zaganiaris
ISBN : 978-2-3221-2844-0
Dépôt légal:août 2019

In loving memory of my wife

Preface

The first application of ion exchange was water softening, introduced in 1905 by the German chemist Gans, who used cation exchange materials called zeolites to exchange Ca^{2+} and Mg^{2+} present in the water for Na^+ of the zeolites. The success of this process encouraged the development of ion exchange resins being chemically stable in acids and alkalis and which could be used to demineralize water. Synthetic ion exchange resins having high capacity and stability based on styrene-divinylbenzene matrix were introduced in 1944. Within a few years, resin manufacturers developed ion exchange resins of strong and weak acid and strong and weak base types with which many "special applications" were developed. Uses such as metals recovery, pharmaceuticals processing, catalysis and food processing were developed as soon as in the late 1940's.

Sugar and corn syrups processing received a special attention because it represented at that time the greatest resin volumes after water treatment, however, many things have changed in the technology used at that time.

Today, food processing has set new challenging objectives such as cost reduction, health considerations, purification and recovery of special products categorized as nutraceuticals, reduced costs of wastes disposal and better managing of the by-products. New applications have been developed like chromatographic separations of carbohydrates, proteins and peptides, recovery of value added polyphenolic compounds and recovery of enzymes used in food processing. In parallel, resin manufacturers have developed products better suited for these applications and engineering companies developed equipment and processes to achieve more efficiently the above mentioned tasks in the food industry.

This book reviews the use of ion exchange resins and synthetic adsorbents in food industries from the point of view of ion exchange rather than of food processing. It results from my experience working in the Ion Exchange Resins department of Rohm and Haas Company and from the numerous visits to relevant companies while in the employ of Rohm and Haas, where discussing with engineers, scientists and technicians brought a lot of understanding and experience. The book is thereforeaddressed to all those working in food processing industries or in parallel industries for whom ion exchange is not their primary field of expertise.

Emmanuel J. Zaganiaris

Acknowledgements

Acknowledgement is made here to Dr Hartmut Haverland and to Michael Latz of Bucher UnipektinAG,toJoseph Fockedey ofCosucra Groupe Warcoing, to Coralie Sowa of Novasepand to DrCraig Jensen of Tongaat-Hulett Sugar Ltd for kindly supplying useful information and other material and permission to use them in this book.
I would like to express my thanks to Osmar Cunha of The Dow Chemical Company in Brazil for helpful information.

ABBREVIATIONS

Å	Angström
ALA	alpha-lactalbumin
BV	Bed Volume
DMA	Dimethylamine
DMF	Dimethylformamide
DVB	Divinylbenzene
CIX	Continuous ion exchange
CSEP	Chromatographic Separation
DE	Dextrose Equivalents
DS	Dry Substances (or Dry Solids)
ED	Electrodialysis
FOS	Fructooligosaccharides
GAC	Granular Activated Carbon
FCOJ	Frozen concentrated orange juice
FFA	Free Fatty Acids
HA	Haze Active
HFCS	High Fructose Corn Syrup
HFS	High Fructose Syrup
HMF	Hydroxymethylfurfural
HMO	Human milk oligosaccharides
HSH	Hydrogenated Starch Hydrolysates
HT	Hydroxytyrosol
ICU	International Color Units
IDA	Iminodiacetic
IEC	Ion Exchange Chromatography
IER	Ion exchange resins
IEZ	Ion exchange zone
IP	Isophytol

ISEP	Ion Separation
ISMB	Improved Simulated Moving Bed
IX	Ion exchange
KHT	Potassium Hydrogen Tartrate
LF	Lactoferrin
LP	Lactoperoxidase
L_R	Liter of resin
MHC	Moisture holding capacity
MP	Macroporous
MR	Macroreticular
MSG	Monosodium glutamate
NF	Nanofiltration
NFC	Not-from-concntrated (juice)
NRS	New Regeneration System
NS	Non-sugars
OMWW	Olive Mill Waste Waters
PVPP	Polyvinylpolypyrrolidone
PW	Pulp Wash
R	Recovery (sugar)
RDN	ResinDioN
SAC	Strong acid cation
SBA	Strong base anion
SMB	Simulated Moving Bed
SSMB	Sequential Simulated Moving Bed
TMHQ	Trimethylhydroquinone
UF	Ultrafiltration
WAC	Weak acid cation
WBA	Weak base anion
WPC	Whey Protein Concentrates
WPH	Whey Protein Hydrolysates
WPI	Whey Protein Isolates

CONTENTS

1. Introduction

Ion exchange is a well known unit operation in many food processing industries. The processes involved are ion exchange, adsorption and chromatography, in column or, more rarely, in batch operations in order to achieve purification, recovery or separations. Ion exchange resins (IER) and polymeric adsorbents are used to achieve various tasks:

- softening
- demineralization
- decolorization
- deacidification
- purification by selectively removing unwanted components
- heterogeneous catalysis
- supports for enzyme immobilization
- chromatographic separations
- recovery of valued products by adsorption and subsequent elution

Food processes considered here include sugar (sucrose), monosaccharides such as glucose, fructose and tagatose, polyols, oligosaccharides, synthetic sweeteners such as sucralose, fruit juices (orange juice, apple juice, other fruit juices and beverages), milk whey, amino acids, organic acids (citric, lactic, succinic or malic acid), gelatin, glycerin, nutraceuticals (vitamins, polyphenols) and various others such as pectins and wine stabilization. It also includes drinking water purification from toxic contaminants.

In the applications of ion exchange resins for food or other "special" applications, the operating conditions are often different from those used in water treatment. In these applications, IER do not only remove ions by ion exchange but they remove also proteins, color material as well as other molecules such as color precursors, off-taste and odor molecules. The removal of these compounds implies not only ion exchange mechanisms but also van der Waals forces, hydrogen bonding or π-bonding interactions. It may also involve a deposition of high molecular weight molecules or aggregates on the resin bead surface or inside the resin which are difficult to remove during the regeneration step of the resin operation. This may result in a premature resin fouling more frequently seen in food applications than in classical water treatment applications so that the life expectancy of the resins may consequently be shorter.

Another difference in resin operating conditions between food and water treatment is the solution concentrations. For example, sugar solutions with concentrations 60% or higher may be treated by ion exchange. Under these conditions, due to osmotic pressure difference, the water content of IER is less than that of the resins surrounded by pure water or dilute solutions.In that case, diffusion of the ions or molecules entering the resin beads is slower compared to that in dilute solutions. In addition, changes from high to low concentrations going from service to sweetening off and on steps provokes a swelling-shrinking cycling that may physically damage the resins. Also, high pressure drops may be experienced due to the high viscosity of the treated solutions caused by high concentrations or by high molecular weight of the solutes as for example in treating syrups or pectin solutions.

In the following chapter, the properties of ion exchange resins and adsorbents are briefly reviewed, including ion exchange resins used as heterogeneous catalysts and as supports for enzyme immobilization. Then the main ion exchange processes are described which include batch operations, column operations and elution chromatography. These items are reviewed only to the extent that allows a better understanding of the use of ion exchange resins and adsorbents as part of the food processing. For a more full understanding of ion exchange resins and adsorbents there are many excellent books and articles that the reader could use. Some references are: Kunin, 1958; Helfferich, 1962; Arden, 1968; de Dardel and Arden, 1989; Clifford, 1999; Kammerer *et al*, 2011.

2. Ion exchange resins and synthetic adsorbents

Ion exchange resins

Structure and properties

Synthetic ion exchange resins, or simply "resins" when the "ion exchange" is implied, are crosslinked polyelectrolytes. Ionic functional groups are attached by covalent bonds to a polymeric matrix. To preserve electro neutrality, an equal number of mobile ions of opposite sign, called counter-ions, are also present. Most commercial resins come in spherical beads, either with polydisperse, gaussian, particle size distribution, having a diameter between 0.3 and 1.2 mm, or with a monodisperse particle size distribution. For certain systems, they may come with special particle size cuts to be compatible with restrictions due to nozzle sizes, pressure drop limitations, fluidization specifications in fluidized beds etc.

From the physical structure point of view, there exist two types of resins. In the first, the IER beads are homogeneous and are called gel-type IER. In the second, designated as macroreticular (MR) or macroporous (MP) IER, the resin beads are heterogeneous and consist of interconnected macropores surrounded by gel-type microbeads agglomerated together (Kun and Kunin, 1967). The macropores have sizes[1] ranging from several angströms (Å) up to many hundreds of angströms while the microbeads give to the resin a large internal surface area which depends on the size of these microbeads. A bead of a gel-type IER is illustrated in figure 2.1 while figure 2.2 illustrates a bead of a macroreticular IER.

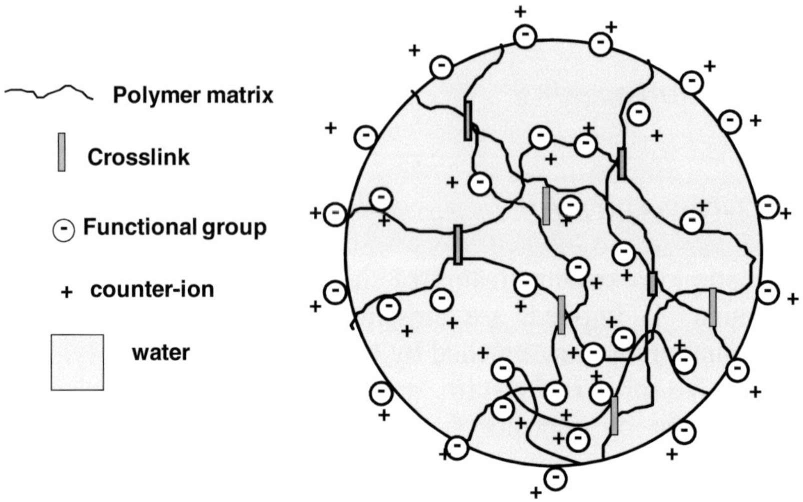

Figure 2.1. Schematic representation of a gel type IER bead

In the case of the resin of figure 2.1, the fixed ionic groups have a negative sign while the counter-ions have a positive sign. Since the mobile, counter-ions are cations, the resin is a cation exchanger. When the fixed ionic groups have a positive sign and

---

[1] By pore size of the macroreticular resins it is meant here the pore diameter, assuming a cylindrical shape of the pores which of course is only an approximation.

22

the counter-ions a negative sign, the resin is an anion exchanger. The ionic groups in the resin may be strong electrolytes or weak electrolytes. There exist therefore strong acid cation (SAC) exchange resins and weak acid cation (WAC) exchange resins. Similarly, there exist strong base anion (SBA) exchange resins and weak base anion (WBA) exchange resins.

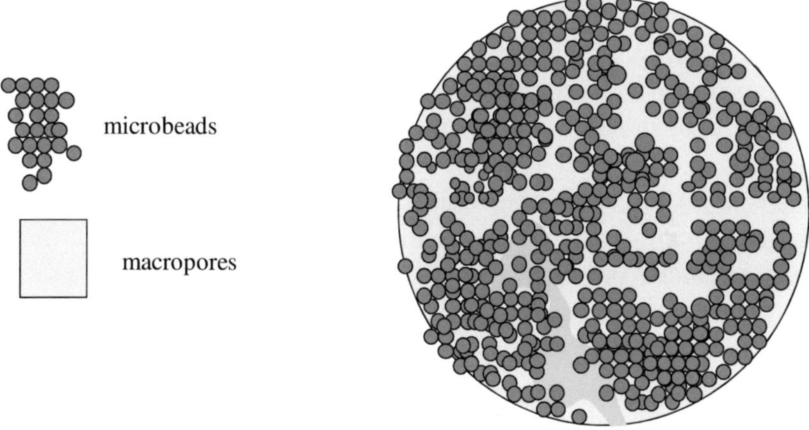

Figure 2.2. Schematic representation of a macroreticular IER bead

The most frequent fixed ionic groups in the commercial IER are sulfonic groups for the SAC resins, carboxylic acid groups for the WAC resins, quaternary ammonium groups for the SBA resins and tertiary ammonium groups for the WBA resins. The polymeric matrix can be a styrene-divinylbenzene (DVB) copolymer or an acrylic-DVB copolymer or a condensation polymer of phenol and formaldehyde. As illustrated in figure 2.1, the IER beads are imbibed with a solvent, usually water. This water is of a primary importance since all the ion exchange reactions take place in solution. The water in the IER can be distinguished as the water in the hydration shells of the counter-ions and the fixed ionic groups and as "free" water. It is through this "free" water that ions or molecules diffuse through the resin beads and therefore it affects strongly the kinetics of ion exchange.

If a dry or partially dry resin comes in contact with water, the water from outside the resin beads will enter into the resin and the resin will swell (Helfferich, 1962). This swelling results from the tendency of the hydrophilic fixed and mobile ionic groups to become hydrated making the polymer chains to expand. Also, since the ionic concentration inside the resin is higher than the ionic concentration in the external solution, water enters into the resin to balance the osmotic pressure difference between the interior of the resin and the external water. A third reason for the swelling of the resin is the electrostatic repulsion of fixed neighbouring ionic groups. The swollen resin will approach equilibrium with the outside water as the above stretching forces (hydration of ions, osmotic pressure difference and the increasing distance of neighbouring ionic groups) decrease. The water contained in a resin in equilibrium with pure water is called moisture holding capacity (MHC), or water uptake or other similar term, and is one of the principal characteristic properties of the resin. The MHC of a resin depends upon the crosslinking density of the polymer matrix, the nature of the polymer (aromatic, acrylic, formophenolic), the nature and the concentration of the ionic functional groups (the total exchange capacity) and the nature of the counter-ions. The effect of the nature of the counter-ions on the degree of swelling is somewhat complex. In general, the resin expands when one counter-ion is replaced by another one of bigger size in the hydrated form. For example, the size of the hydrated cations varies in the following order:

$$H^+ > Li^+ > Na^+ > K^+ > Rb^+ > Cs^+$$

A sulfonic SAC resin therefore swells going from the Na^+ form to the H^+ form and shrinks going from the Na^+ to the K^+ form. However, as the bigger counter-ion takes the place of a smaller one, the polymer matrix will apply a higher swelling pressure and even though the resin will expand, it will expel some of the free water. The higher the crosslinking density, the higher will be the internal swelling pressure and more "free" water will be expelled. The swelling therefore from one ionic form to another will depend on the crosslinking density. For example, a 8% DVB SAC resin swells by 16% going from the K^+ to the H^+ form while a 10% DVB resin swells by 9%. For resins with low crosslinking density where the free water represents a big part of the moisture content of the resin, the effect of the counter-ion valence on swelling becomes significant. In fact, it is the number of hydrated counter-ions that contributes to the moisture content. Since on the same equivalent basis, the number of divalent for example ions is half of those of monovalent, by converting a low crosslinked resin from the monovalent to the divalent form, the resin shrinks. With highly crosslinked resins, this may be reversed because there is not much free water and the divalent ions may be highly hydrated so that even with half the number, the resin may swell going from mono- to divalent form (Helfferich, 1962).

When an IER comes in contact with an aqueous solution rather than with pure water, the resin swells less than in pure water because the osmotic pressure difference between the resin and the external solution is smaller. If the external solution is very concentrated, it can be that the resin shrinks as water is displaced from the resin to the external solution. This is seen for example when the resins are regenerated with concentrated acids or caustic solutions. It is also seen when resins treat concentrated sugar syrups. In these cases the "free" water is considerably reduced and the kinetics of ion exchange is reduced as well. Also, the properties that characterize ion exchange resins, such as moisture content, true wet density, expansion and pressure drop curves or swelling from one ionic form to another, as usually provided by the resin manufacturers in product data sheets, refer to resins in equilibrium with pure water. When the resin is in equilibrium with a solution, then these properties may have different values.

If a SAC resin in A^+ form comes in contact with a dilute solution containing a strong electrolyte A^+Y^- then the cation A^+ will be found at higher concentration inside the resin beads than in the external solution while the anion Y^- (co-ion) will be found in higher concentration in the external solution. There will be then a beginning of migration of cations A^+ towards the external solution and of anions Y^- towards the resin. This however will create a net positive charge in the external solution and a net negative charge inside the resin beads building up an electric potential difference between the two phases. This so called Donnan potential will prevent anions Y^- from entering the resin beads and cations A^+ from coming out of the resin. Because of electro neutrality requirement, the co-ion Y^- exclusion from the resin means electrolyte A^+Y^- exclusion. This exclusion is called Donnan exclusion (Helfferich, 1962).

As seen above, ion exchangers are poor sorbents for strong electrolytes. However, they are good sorbents for nonelectrolytes. From a mixture of electrolytes and nonelectrolytes, the resin excludes the strong electrolytes and takes up the nonelectrolytes. This ion exclusion phenomenon is used to separate neutral molecules such as sugar, glycerin (Prielipp and Keller, 1955; Asher and Simpson, 1956; Diaion, 1995) etc from mineral salts. Ion exclusion chromatography is discussed in the next chapter.

Ion exchange equilibrium

Take a SAC exchange resin and place it in a receipient containing a solution of an electrolyte. If the electrolyte in the external solution contains a different counter-ion from that of the ion exchanger, then ion exchange takes place between resin and solution until an equilibrium is attained, where the composition of the external solution and of the resin remain constant.

Consider the case where rhe SAC resin is in the H^+ form (that is, the counter ion is a H^+) that comes in contact with an aqueous solution of NaCl. An exchange takes place of the H^+ for the Na^+. After equilibrium has been established, the IER, initially in the H^+ form, will be found in a form partially H^+ and partially Na^+. Similarly, the solution, initially NaCl, will contain both, NaCl and HCl. The ion exchange reaction is described by eq. (2.1):

$$\mathcal{R}\text{-}H + Na^+Cl^- \rightleftarrows \mathcal{R}\text{-}Na + H^+Cl^- \qquad (2.1)$$

The equilibrium condition is described with the equilibrium constant (2.2):

$$K_{Na/H} = \frac{(R\text{-}Na)(H^+)}{(R\text{-}H)(Na^+)} = \frac{q_{Na}\, c_H}{q_H\, c_{Na}} \qquad (2.2)$$

where $K_{Na/H}$ is the selectivity coefficient and ( ) denote concentrations in equivalents per liter (eq/L) or equivalents per liter resin (eq/L$_R$). c_i and q_i denotes concentrations in solution and on the resin respectively[2].

Alternatively, the use of the separation factor, $\alpha_{i/j}$, defined as:

$$\alpha_{i/j} \;=\; \frac{\text{distribution of ion i between phases}}{\text{distribution of ion j between phases}} = \frac{y_i \,/\, x_i}{y_j \,/\, x_j} \qquad (2.3)$$

is more practical. Here :

x_i is the equivalent fraction of ion i in solution and

---

[2] Equivalent is the number of moles of an ion multiplied by the valence of that ion. The equivalent weight is therefore equal to the molecular weight devided by the valence of the ion. The concentration in equivalents (eq) or milliequivalents (meq) per liter is equal to the mass devided by the equivalent weight. For example, the Ca^{2+} ion has an atomic weight of 40, the equivalent weight is then 40/2=20 and a solution of 1 g Ca^{2+}/L is 1/20=0.05 eq/L or 50 meq/L. In ion exchange, because the various ions have different valence, the concentrations in eq/L or meq/L are widely used.

y_i is the equivalent fraction of ion i in the resin.

$$x_i = c_i / C \quad \text{and} \quad y_i = q_i / Q \qquad (2.4)$$

Where c_i is the concentration of ion i in solution in eq/L and C the total concentration of all ions in solution in eq/L. q_i is the concentration of ion i on the resin in eq/L_R and Q is the total capacity of the resin in eq/L_R.
If the resin prefers ion i over ion j we call this a favorable equilibrium and $a_{i/j} > 1$. In the opposite case we call this an unfavorable equilibrium and $a_{i/j} < 1$.

Using eqs (2.2) and (2.3) we have for the above example of the H^+/Na^+ exchange :

$$\alpha_{Na/H} = \frac{q_{Na}\, c_H}{q_H\, c_{Na}} = K_{Na/H} \qquad (2.5)$$

The separation factor α is equal to the selectivity coefficient K.

From the equilibrium concentrations of the ions in the resin and the equilibrium concentrations in solution, at constant temperature, one can obtain the value of $a_{Na/H}$ or of $K_{Na/H}$. The plot, y_i <u>vs</u> x_i, is called ion exchange isotherm.

Consider the case of softening using a SAC resin in the Na^+ form. The exchange reaction is:

$$2\,\mathcal{R}\text{-Na} + Ca^{2+} \rightleftharpoons \mathcal{R}_2\text{-Ca} + 2\,Na^+ \qquad (2.6)$$

where $\mathcal{R}$ represents the polymer matrix of the resin. The equilibrium condition is described with the equilibrium constant (2.2):

$$K_{Ca/Na} = \frac{(\mathcal{R}_2\text{-}Ca)(Na^+)^2}{(\mathcal{R}\text{-}Na)^2(Ca^{2+})} = \frac{q_{Ca}\, c_{Na}^2}{q_{Na}^2\, c_{Ca}} \qquad (2.7)$$

where $K_{Ca/Na}$ is the selectivity coefficient and ( ) denote concentrations in equivalents per liter (eq/L) or equivalents per liter resin (eq/L$_R$) and where

$$q_{Ca} = \text{concentration of Ca in resin in eq/L}_R$$
$$c_{Na} = \text{concentration of Na in solution in eq/L}$$
$$q_{Na} = \text{concentration of Na in resin in eq/L}_R$$
$$c_{Ca} = \text{concentration of Ca in solution in eq/L}$$

Using the separation factor α we have:

$$\alpha_{Ca/Na} = \frac{q_{Ca}\, c_{Na}}{q_{Na}\, c_{Ca}} = \frac{y_{Ca} x_{Na}}{y_{Na} x_{Ca}} \qquad (2.8)$$

where x is the equivalent fraction of Na or Ca in solution and y is the equivalent fraction of these ions in the resin:

$$x_{Na} = c_{Na}/C \quad \text{and} \quad y_{Na} = q_{Na}/Q \quad \text{for } Na^+ \text{ and}$$

$$x_{Ca} = c_{Ca}/C \quad \text{and} \quad y_{Ca} = q_{Ca}/Q \quad \text{for } Ca^{2+}$$

where C is the total concentration of all ions in solution in eq/L and Q is the total capacity of the resin in equivalents per liter of resin (eq/L$_R$).

As seen from eq. (2.8), even if the resin has higher affinity for Ca^{2+} over Na^+, in other words $\alpha_{Ca/Na}$ is greater than 1, the higher is the Na^+ / Ca^{2+} concentration ratio in solution, the smaller will be the concentration ratio of Ca^{2+} / Na^+ in the resin. A typical thin juice composition has a Na^+ plus K^+ concentration in the range of 50-100 meq/L while Ca^{2+} in the range of 3-8 meq/L. At this thin juice composition, at equilibrium, the resin will contain relatively high levels of Na^+. Taking a selectivity coefficient $K_{Ca/Na}$ of 2.6, at equilibrium, about 45% of the resin will be in the Na^+ form and 55% in the Ca^{2+} form while with many domestic water compositions, the resin, at equilibrium, would be for more than 80% in the Ca^{2+} form. Therefore, the operating capacity of the resins in thin juice decalcification cannot be very high due to the thin juice composition and not due to an incomplete regeneration for example. Usually, operating capacities in thin juice decalcification are in the range of 0.5 to 0.8 eq/L_R while in domestic water softening they exceed the 1 eq/L_R.

From eq. 2.7 and 2.8 one obtains:

$$\alpha_{Ca/Na} = K_{Ca/Na}\,(q_{Na} / c_{Ca}) \qquad (2.9)$$

Therefore, the separation factor when the valence of the two ions in equilibrium is not the same, is not constant but it depends on the composition of the solution and the resin. Depending on the solution and resin composition, the separation factor can be either >1 (favourable equilibrium) or <1 (unfavourable equilibrium) for the same value of the selectivity coefficient $K_{Ca/Na}$. Using fractions rather than concentrations eq. (2.7) becomes

$$\frac{y_{Ca}}{(1-y_{Ca})^2} = K_{Ca/Na} \; \frac{Q}{C} \; \frac{x_{Ca}}{(1-x_{Ca})^2} \qquad (2.10)$$

From eq (2.10) it is seen that the isotherm would depend on the total ionic concentration in solution, C, since the total resin capacity, Q, is constant. Figure 2.3 illustrates the case for two different values of C, using eq. (2.10). The curve with C=0.01 eq/L represents loading conditions while that with C=2 eq/L regeneration conditions.

Going back to the softening case using a SAC resin in the Na^+ form discussed above, the regeneration reaction is the reverse of that of the loading reaction (2.6):

$$\mathcal{R}_2\text{-Ca} + 2\,Na^+ \; \leftrightarrows \; 2\,\mathcal{R}\text{-Na} + Ca^{2+}$$

Even though during loading and therefore at relatively low ionic concentration, the affinity of the resin is higher for Ca^{2+} than for Na^+, as seen in figure 2.3, during regeneration at high regenerant concentration, the affinity of the resin for Ca^{2+} is significantly lower compared to Na^+ and can even be reversed. Therefore, a SAC resin fixes favourably Ca^{2+} during loading but during regeneration, Ca^{2+} is eluted off the resin without too much difficulty.

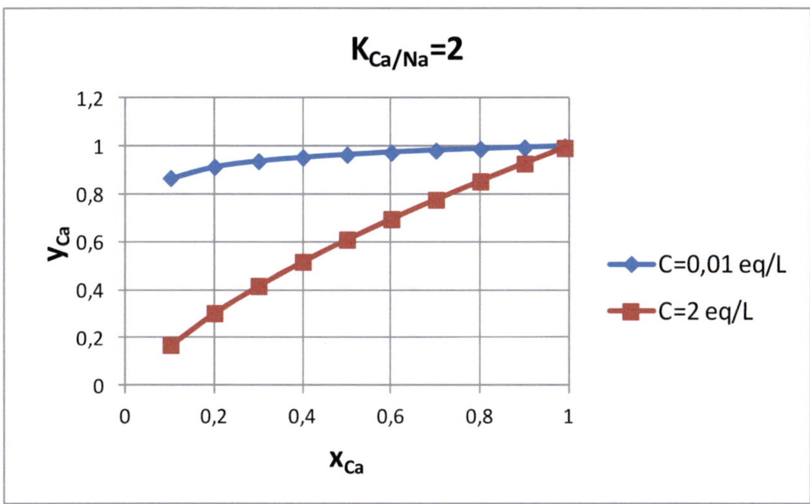

Figure 2.3 Ion exchange isotherms for Ca/Na equilibrium with $K_{Ca^{2+}/Na^{+}} = 2$. Curve with total solution concentration of 0.01 eq/L is favourable, curve with total solution concentration of 2.0 eq/L is less favourable

This simple picture of the decalcification reaction can become more complex if in solution there are other ions that interact with Ca^{2+} and thus compete with the functional groups of the ion exchange resin, $\mathcal{R}\text{-}SO_3^-$. For example, if there are citrate, oxalate or phosphate ions in solution, there is formation of complexes with Ca^{2+}, so that the Ca^{2+} ions are distributed between these complexes and the functional groups of the resin:

$$(citr) + Ca^{2+} \;\rightleftharpoons\; (citr)\text{-}Ca^{2+} + 2\,H^{+} \qquad K_1$$

$$2\,\mathcal{R}\text{-}SO_3^- Na^{+} + Ca^{2+} \;\rightleftharpoons\; (\mathcal{R}\text{-}SO_3^-\,)_2 Ca^{2+} + 2\,Na^{+} \quad K_2$$

The ratio of Ca^{2+} fixed on the resin to that remaining in solution by forming the complex depends therefore on the values of K_1 and K_2 and on the concentration of citrates and the resin functional groups in solution.

This situation is presented in some juices where there in no carbonatation treatment which removes the complexing anions or in other cases like milk whey where these complexing anions exist in the mother liquors. These situations have been faced by treating the solutions on a strong base resin in the Cl form, before the decalcification cation exchanger, thus removing the complexing anions. Regeneration is done with NaCl, in series SAC ⇨SBA resin.

In multicomponent systems, we can establish a selectivity sequence by comparing the separation factors of various ions with respect to one ion taken as a reference. Thus, for a SAC resin we have the following sequence of selectivities taking the H^+ as reference (de Dardel and Arden, 1989) :

$$Ca^{2+} > Cu^{2+} > Zn^{2+} > Mg^{2+} > K^+ > NH_4^+ > Na^+ > H^+$$

Similar considerations apply to SBA resins. Since the resin is a strong electrolyte and is dissociated 100%, during ion exchange is always an equilibrium that is established between ions in the external solution and on the resin. The selectivity sequence for SBA resins is as follows:

$$SO_4^{2-} > CO_3^{2-} > HSO_4^- > NO_3^- > Cl^- > HCO_3^- > OH^-$$

Weak electrolyte resins due to their partial dissociation, behave differently than strong electrolyte resins.

Take a WAC resin bearing a carboxylic acid group and consider the case where this carboxylic resin is placed in contact with a solution containing a salt M^+X^-. The exchange reactions are as follows:

$$\mathcal{R}COOH \rightleftharpoons \mathcal{R}\text{-}COO^- + H^+ \qquad (2.11)$$

$$\mathcal{R}\text{-}COO^- H^+ + M^+X^- \rightleftharpoons \mathcal{R}\text{-}COO^-M^+ + HX \qquad (2.12)$$

$$HX \rightleftharpoons H^+ + X^- \qquad (2.13)$$

If acid HX is stronger than the carboxylic acid group of the resin, then reaction (2.12) will go to the left and the resin will not significantly fix M^+ ions. In this case we say that the carboxylic resin does not salt-split. In order that reaction (2.12) goes to the right, the acid HX should be weaker than the carboxylic group, or that the dissociation constant Ka (resin) is higher than the dissociation constant Ka (acid HX), or

$$pK_a \text{ (resin)} < pK_a \text{ (acid HX)} \qquad (2.14)$$

Polyacrylic acid has a pK_a of about 5 while polymethacrylic acid has a pK_a of about 6. Carbonic acid has a pK_{a1} of 6.34. This means that a polyacrylic acid resin in the H form can fix Na^+ from a $NaHCO_3$ solution and replace it with H^+. Since however the two pK_a values, that of the resin and that of the carbonic acid, are not very different, the resin will fix only partially Na^+ in exchange for H^+ and certain part will remain in the H form.

It should be noted that cation exchangers with the carboxylic functional groups show the opposite affinity series for alkali and alkaline earth metal ions (Gregor *et al*, 1956) compared to SAC resins. This is due to the fact that the carboxylic group prefers larger, more hydrated and more polarizable counter-ions, the contrary of sulfonic resins. The affinity of this type of cations for carboxylic resins is therefore as follows:

$H+ > Mg2+ > Ca2+ > Sr2+ > Ba2+ > Li+ > Na+ > K+ > Rb+ > Cs+$.

WBA exchangers react in a similar way:

$$\mathcal{R}\text{-}CH_2N(CH_3)_2 + H_2O \rightleftarrows \mathcal{R}\text{-}CH_2NH^+(CH_3)_2\ OH^- \quad (2.15)$$

$$\mathcal{R}\text{-}CH_2NH^+(CH_3)_2\ OH^- + M^+X \rightleftarrows \mathcal{R}\text{-}CH_2NH^+(CH_3)_2X^- + MOH \quad (2.16)$$

$$MOH \rightleftarrows M^+ + OH^- \quad (2.17)$$

If the base MOH is a strong base, then in a batch operation, reaction (2.16) will go to the left and the WBA resin will not fix the anion X^-, in other words, the WBA resin does not salt-split.

To be more precise, weak electrolyte resins, WAC and WBA, even at neutral pH have a certain salt-splitting capaciry. A WBA for example, gives:

$$R\text{-}NR_1R_2 + H_2O \rightleftarrows R\text{-}NH^+R_1R_2 + OH-$$

with a dissociation constant, $\quad K_b = \dfrac{(R\text{-}NH+)(OH-)}{(RN\ R_1R_2)} \quad (2.18)$

$$R\text{-}NH^+R_1R_2 \ + \ NaCl \ \leftrightarrows \ R\text{-}NH^+R_1R_2 \ Cl^- \ + \ NaOH \quad (2.19)$$

In a column operation, the NaOH formed is carried away, at least partially, with the effluent.

A styrenic WBA resin has pK_b values about 6, an acrylic WBA has a pK_b about 5. Using eq (2.18) it can be calculated a degree of ionization of 0.09% for a styrenic and 0.25% ionization for an acrylic WBA resin. This is however very low degree of ionization and with a typical water containing several meq/L of salts, the cycle length of the resin will be very short and we say that weak electrolyze resins do not split salts. This can change when only traces of a salt have to be removed, for which the resin has high affinity, in presence of other salts in higher concentrations but for which the resin has significantly lower affinity.

Another characteristic feature of the WBA resins is that they do not fix very weak acids.

$$\mathcal{R}\text{-}CH_2N(CH_3)_2 \ + \ HX \ \leftrightarrows \quad \mathcal{R}\text{-}CH_2NH^+(CH_3)_2 + \ X^- \quad (2.20)$$

If the acid HX is weaker acid than the resin, or if

$$pKa \ (HX) > pK_a \ (resin) \quad\quad (2.21)$$

then the reaction (2.20) will go to the left. For example, WBA resins do not fix very weak acids such as silicic acid, H_2SiO_3, which has a pK_{a1} of 10, while styrenic WBA resins have a pK_a of about 8.5 and acrylic weak base resins a pK_a of about 9.

Both, WAC resins and WBA resins regenerate quantitatively with only a small excess of regenerant since even a small excess of acid for the WAC resins or of a base for the WBA resins will suppress the dissociation of the resins and therefore the resin will release any ions that may be fixed on the functional groups. In addition, WBA resins, contrary to the SBA resins, can be regenerated with weak bases like NH_4OH or Na_2CO_3. Similarly, WAC of mathacrylic acid type can be regenerated with CO_2 dissolved in water under pressure (H_2CO_3).

WAC resins in the H form and WBA resins in the free base form are hydrophobic while these groups in the dissociated form are hydrophilic. For that reason, when WAC or WBA resins go from the regenerated form to the exhausted (dissociated) form swell considerably and this should be taken into account in designing a unit with these resins or defining the operating conditions (for example including a short backwash to release any pressure developed by a compacted resin bed).

Most of the styrene-DVB WBA exchange resins in the commerce have strong base groups at a level of about 10-15% of the total exchange capacity resulting from side reactions during manufacturing (Zaganiaris, 2017) and which have a significant effect on the properties of the WBA resins. Since the strong base groups are fully dissociated, the WBA resins have a certain hydrophilicity due to these groups thus accelerating the kinetics of exchange at the beginning of the cycle. Also, the presence of strong base groups result into a lower swelling due to the fact that the strong base groups shrink going from the regenerated (OH^-) form to the exhausted form while the weak base groups swell. The overall swelling of the resin going from the regenerated (OH^- or free base) form to the exhausted form is less than it would be without strong base groups. In addition, these strong base groups can split salts or they can fix very weak acids

as long as they are found in the OH⁻ form. It should also be added that the strong base capacity give a higher pH inside the resin beads during the beginning of the cycle when these groups are still in the OH⁻ form. This has an impact if the treated solution contains compounds that are sensitive to pH (for example glucose or fructose).
Acrylic WBA resins have no strong base functionality.

Ligand exchange

In many of the food processes like drinking water purifications or glucose-fructose separation presented in this book, the mechanism involved is ligand exchange, a mechanism first introduced by Helfferich (Helfferich, 1961). This section is a short reminder of this mechanism.
A metal ion can form coordination complexes with neutral molecules or anions, called ligands, via interactions between lone pairs of electrons of the outer energy level of the ligand and the s, p and d orbitals of the metal. The ligand is a Lewis base (donates electrons) and the metal is a Lewis acid (accepts electrons). The bond between the ligand and the metal, where the ligand donates the electron pair, is called coordination bond, as opposed to a covalent bond where each atom provides one electron. Generally speaking, the strength of a coordination bond is in between that of a covalent bond and an electrostatic attraction bond in ion exchange.

$$\text{L:} \;\; + \;\; M^{2+} \;\; \leftrightarrows \;\; L{\rightarrow}M^{2+}$$

$$\text{Ligand} \qquad \text{Metal} \qquad \text{Coordination complex}$$

The arrow denotes the coordination bond.
Examples of common ligands are: Cl^-, NH_3, H_2O, OH^-, O^{2-}, CN^-, represented by the dot Lewis structure by:

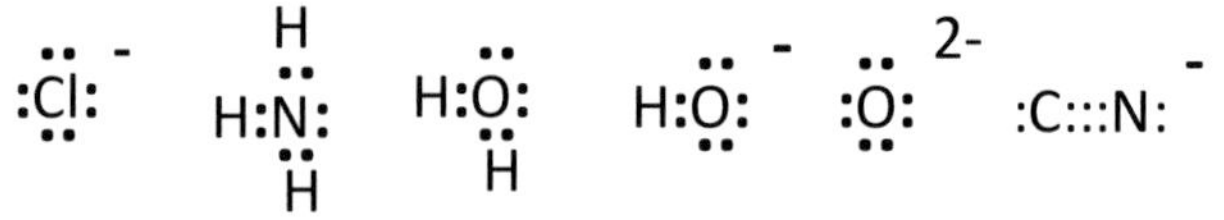

If in addition, the ligand has anionic groups and is charged negatively, the complex is even more stable because the metal is fixed by coordination bond and electrostatic attraction at the same time. The classical example is the iminodiacetic (IDA) resins bearing two negative charges.

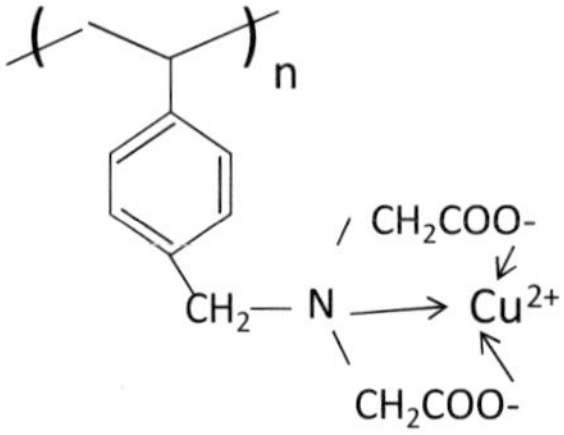

Here, the ligand is also the functional group of the resin. This type of resins is called *selective resins* or *chelating resins*.

The central metal atom is loaded with ligands like H_2O, NH_3, Cl^-, OH^-, or CN^-. for example the copper (II) aquo complex $[Cu(H_2O)_6]^{2+}$:

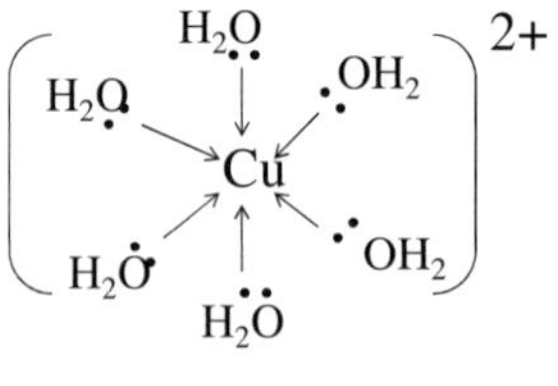

When the Cu^{2+} atom is fixed on the resin by coordination bond, as in the case of IDA resins, then three of the water molecules are replaced by the three bonds, one to the N atom and two to the -C(O)O$^-$ groups. The Cu^{2+} remains then with three water molecules.

Ligand exchange is then the exchange of water ligands of the central metal atom by another ligand found in the water solution:

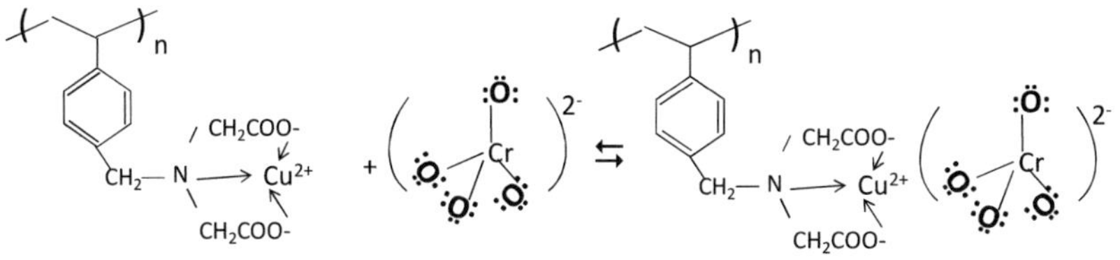

If the ligand in the water solution is negatively charged, then as mentioned above, the complex formed is more stable because this ligand is bound to the metal by a coordination bond and in addition is attracted by electrostatic forces.

Ligand exchange resins allow the selective removal of ligand impurities, such as fluorides, chromates, phosphates, arsenates and others, from water even in presence of higher concentrations of common anions like Cl^-, SO_4^{2-}, NO_3^- or HCO_3^- (Sarkal *et al*, 2016).

Ion exchange kinetics

The rate at which the exchange of ions reaches equilibrium is not instantaneous. It depends upon the mass transfer of the ion i in solution to the exchange sites in the resin and that of the ion j from the exchange sites into the solution. This mass transfer of ions from the solution to the resin and vice versa is essentially a diffusion process (Helfferich 1966).

Around a resin bead in contact with a solution, there is a stagnant layer of liquid called Nernst film. Consider a SAC resin in the H^+ form in contact with a solution of NaCl. Assuming that the transfer of ion Na^+ from the bulk of the solution to the Nernst film is instantaneous due to the agitation of the solution, the overall rate of ion exchange consists of the rate of the diffusion of Na^+ through the Nernst film and the diffusion through the resin particle. If the rate of diffusion through the Nernst film is slow compared to the diffusion through the resin, the kinetics are film diffusion controlled. If on the contrary the diffusion through the resin particles is slow compared to the diffusion through the Nernst film, the kinetics are particle diffusion controlled.

The diffusion of the Na+ ions into the resin particle is characterized by a diffusion coefficient, $\underline{D}_{Na+}$. This diffusion coefficient is slower than the diffusion coefficient D_{Na+} of Na^+ in solution. The reason is that in a resin particle, part of the space is occupied by the polymer matrix while the ions diffuse only through the water phase. There are other reasons as well, for example the path that ions follow is not a straight line but tortuous or the size of the ions in comparison to the size of the openings of the polymer network may be large. For these reasons, some models were developed to predict an effective diffusion coefficient $\underline{D}$ in relation with D. They give expressions varying between

$$\underline{D}=D(\varepsilon/2)$$

and

$$\underline{D}=D[\varepsilon/(2-\varepsilon)]^2$$

where ε is the fractional pore volume of the resin particle.

In order to derive the rate of ion exchange, that is the fractional attainment of equilibrium, as a function of time, there exist two approaches. One is to set up the appropriate flux equations for the diffusing species and integrate them to obtain ion exchange rates. The other is to postulate rate equations involving reaction rates or mass transfer coefficients, which are easier to integrate.

Assume that A is the counter-ion in the resin particleand B in the solution and an exchange is taking place between A and B. Ion A is diffusing out of the resin due to its concentration difference between the resin and the solution. In order to preserve electroneutrality, an equal number of ions B (assuming equal valence) have to diffuse from the solution into the resin. If the mobilities of A and B are not the same, an electrical field will develop which tends to accelerate the slower ion and to slow down the faster one. We have therefore a flux where concentration difference is the driving force (and where Fick's 1st law applies)

$$J_{Ad} = -\underline{D}_A grad \underline{C}_A$$

and a flux where electrical potential gradient is the driving force

$$J_{Ael} = -\underline{D}_A\underline{C}_A(z_A F/RT) grad \phi$$

where F is Faraday's constant, ϕ the electric potential, R the gas constant, T the absolute temperature.
The combined flux is given by the Nernst-Planck equation. From the flux equation one derives the concentration profiles

$$d\underline{C}_A/dt = -div J_A$$

and the reaction rate expressed as the fractional attainment of equilibrium, F(t) as a function of time.

In most cases, the solution of the differential flux equations involves great mathematical difficulties. For that reason, frequently it is preferred to take simplified differential reaction rate equations of the form

$$-dc_A/dt = f(\,c_A,\, c_B,\, q_A,\, q_B,\, \ldots)$$

which are more readily integrated.

The flux of the transfer of Na^+ through the Nernst film follows Fick's law and is characterized by a diffusion coefficient, the concentration difference between the two sides of the film, Δc, and the thickness of the film, δ. Factors that affect the film-controlled kinetics, assuming that the diffusion coefficient is constant, are:

a) Factors that increase the Δc:
- high resin capacity and high affinity for the in-coming ion; these factors tend to keep the ion concentration near the resin surface at low values
- high solution concentration

b) factors that decrease the film thickness:
- high linear velocity of the solution through the column or vigorous agitation of the solution
- small bead diameter

The rate of ion transfer inside the resin particle does not depend on the external conditions of concentration but depends on the particle size of the resin and the internal diffusion coefficients.

The diffusion of ions through the resin beads depends on the structure of the resin. Highly crosslinked polymer phase result in low moisture, low swelling and slow diffusion. Weak electrolyte, WAC and WBA resins, constitute a particular case since in their non-dissociated, H form for the WAC and free base form for the WBA resins, these resins are hydrophobic, they contain low moisture and diffusion is slow. As soon as they are converted to the dissociated form however, they swell considerably and diffusion increases accordingly.

Synthetic adsorbents

Adsorption, like ion exchange and chromatography, is a process through which dissolved atoms, ions or molecules are transferred selectively from the solution to solid particles mixed with the solution and adhere on the surface of the solid particles by weak van der Waals forces. Generally speaking, hydrophobic or non-polar molecules or portions of a molecule, are attracted by hydrophobic, non-polar surfaces while polar molecules are attracted by polar surfaces. Synthetic adsorbents offer a variety of surface polarities, aromatic, acrylic, formophenolic, and can be used in a variety of applications.

Adsorbents are solid materials that offer a surface on which atoms or molecules found in a fluid in contact with the adsorbent can be attracted. This results into a distribution of these atoms and molecules between the adsorbent and the fluid and their selective transfer from the fluid to the surface of the solid adsorbent.

Adsorbents have a high specific surface area (m^2 per unit weight) which is obtained either by grinding a solid to very small particles to form a fine powder, or by creating a porosity inside the adsorbent particles. Porous adsorbents are characterized by a porosity (ml per gram or ml per ml of adsorbent), a specific surface area (m^2 per gram) and a pore size and pore size distribution. Granular carbon, bone char and activated alumina are examples of porous adsorbents obtained from naturally occurring materials. In what follows, only synthetic polymeric adsorbents are considered.

The size of the pores (pore diameter) is classified as micropores, mesopores and macropores, defined as follows:

Micropores <20 Angströms
Mesopores 20 – 500 Angströms
Macropores > 500 Angströms

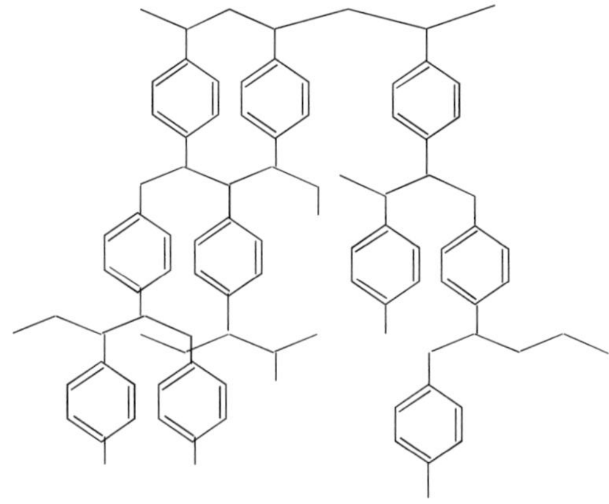

Figure 2.4 styrene-DVB adsorbents

46

Synthetic adsorbents are porous polymeric materials most frequently coming in the form of spherical beads with size between 0.3 and 1.2 mm. They come as styrene-DVB copolymers (fig. 2.4), as acrylic copolymers (fig. 2.5), or as phenol-formaldehyde condensates (fig. 2.6), these last ones coming in granular form. As it is the case with ion exchange resins, synthetic adsorbents can be used in cyclic processes in vessels or in columns.

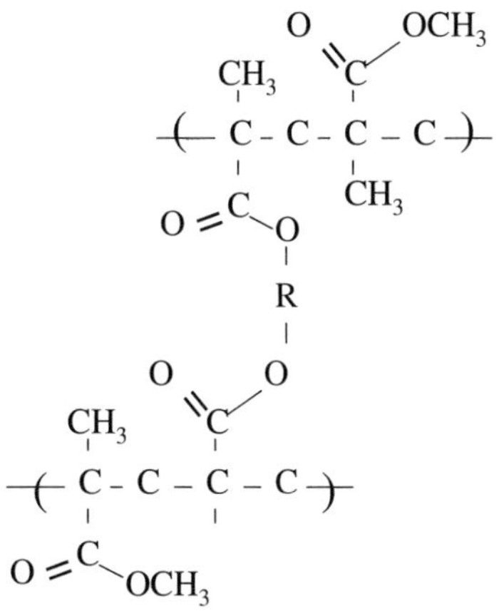

Figure 2.5 Acrylic adsorbent

In principle, aromatic adsorbents come without any functional groups and are hydrophobic materials. They interact with dissolved molecules mainly by π-bond interactions. In addition, there exist aromatic adsorbents bearing functional groups, either sulfonic (cation exchange) or amine (anion exchange) groups.

The purpose of these functional groups is to give some hydrophilicity to the matrix still keeping the aromatic backbone in order to favour the adsorption of certain compounds. This kind of adsorbents having some ionic functional groups can therefore alter the ionic composition of the liquids in which they are used.

Acrylic adsorbents are polymethylmethacrylates crosslinked with a diol or a triol. They are hydrophilic and they interact with dissolved molecules mainly with van der Waals forces or H-bonds. Phenolic adsorbents have the structure of fig.2.6. Phenolic groups being weak acids, these adsorbents have therefore ionic groups and can function as ion exchangers and as adsorbents. They can, under certain conditions of pH, modify the ionic composition of the solution by cation exchange and this should be taken into account when only adsorption is desired without altering the ionic composition of the treated solutions.

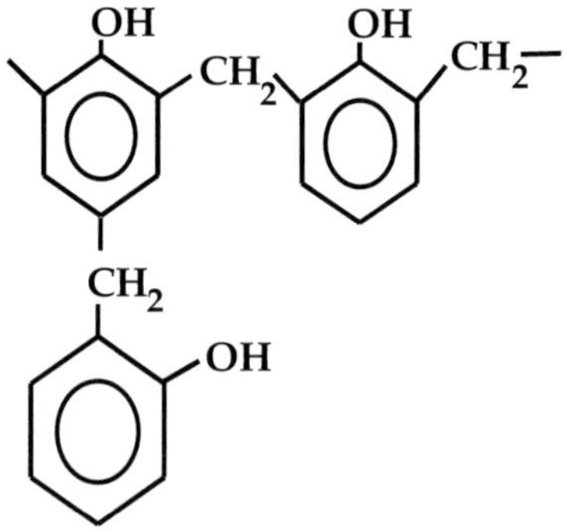

Figure 2.6 Formophenolic adsorbent

48

Synthetic adsorbents were made in the end of the years 1950's by Rohm and Haas Company, Bayer and Permutit and produced industrially in the early 1960's by Rohm and Haas Company. They were made by incorporating into the mixture of monomers a solvent, called porogen, which is a good solvent for the monomers but a non-solvent for the formed polymer which then precipitates out forming microbeads "glued" together. When polymerization is terminated, the porogen is removed leaving a space, the pores, between the microbeads in a way that one bead of a synthetic adsorbent is an agglomerate of microbeads, as it was illustrated in fig. 2.2

By controlling the polymerization conditions, the monomer composition, the initiator concentration, temperature, the quantity and the type of porogen, one obtains a variety of structures having a given surface area, porosity and a certain pore size and pore size distribution. Assuming a certain shape of the pores, for example a cylindrical shape, an average pore diameter can be calculated from the specific surface area (m²/g) and the porosity (ml/g)[3]. In fact, for a given porosity, the size of the pores is inversely proportional to the surface area. It is obvious that the indication of an average pore size given above is only a rough approximation, for example, the average pore size takes into account only the total porosity and the surface area and it does not give the distribution of pore sizes which plays a major role in adsorption capacity and kinetics. Adsorbents having both small and large pores often offer an

---

[3] Assuming a cylindrial geometry of the pores, the average pore diameter can be obtained using the formula :

average pore diameter (in Angströms) = 40000 P/S

  where P is the porosity in ml/g and S the specific surface area in m²/g easier access to the small pores and therefore a better utilization of the internal surface.

With the above technique, adsorbents having a specific surface area in the range of 400 up to about 1000 m^2/g with an average pore diameter from 300 Å down to 50 Å respectively were obtained.

Davankov and co-workers (Davankov and Tsyurupa, 1989) developed a new type of adsorbents where the porosity was introduced not during polymerization but during a post-crosslinking step. To the mixture of the monomers a diluent is used that is a good solvent for the formed polymer. After polymerization, the polymer, either linear or highly swollen polymer is post-crosslinked with an external crosslinking agent where it is obtained a rigid hypercrosslinked structure having a very high surface area (800-1500 m^2/g) and a small average pore diameter, in the range of 10-40 Å. These adsorbents are called Hypersol- Macronet^{TM} sorbents (Purolite International Ltd.).

In 1981, Rohm and Haas Company developed a new type of macronet adsorbents (Reed, 1981). These adsorbents were obtained from a macroreticular lightly crosslinked copolymer beads swollen in an inert solvent and then post-crosslinked using an external crosslinking agent.

An indication of the pore structure of some commercial products is given in Table 1.

Regeneration of the adsorption can take place in various ways:
- by using a different solvent than in the adsorption step whereby the distribution of the adsorbed species between adsorbent and solution is reversed.
- By changing the pH, and hence changing the solubility, of certain ionic adsorbed species, for example by eluting

phenol or phenolic compounds from aromatic adsorbents with NaOH.

- By using steam to desorb volatile adsorbed compounds including those that form azeotropes with water.

Table 1 Representative values of pore structure of synthetic adsorbents

	Macroreticular adsorbents				Macronet adsorbents		
	Acrylic	Aromatic, small pores	Aromatic, medium pores	Aromatic large pores	Aromatic, small and large pores	Aromatic small and large pores	Activated carbon
Micro, ml/g	0.26	0.43	0.45	0.31	0.35	0.44	0.48
Meso, ml/g	0.51	0.86	0.93	0.61	0.13	0.16	0.09
Macro, ml/g	0.33	0.07	0.45	0.92	0.48	0.48	0.04
Total porosity, ml/g	1.10	1.36	1.83	1.84	0.96	1.01	0.61
Spec. Surface area, m²/g	540	900	950	650	800	1020	950
Average pore diameter, Å	82	61	77	113	48	40	26

Adsorption can be modelized by assuming that the adsorbate is attached in a monolayer on the internal surface of the adsorbent (Langmuir model). According to this model, all the sites of the adsorbent are equivalent and can fix only one molecule of the adsorbate. Taking the adsorption of a species A adsorbed on the sites S of the adsorbent, we have:

$$S + A \rightleftharpoons SA$$

with equilibrium constant:

$$K = q/c\,(S) \qquad (2.22)$$

where (S) represents the free sites on the internal surface of the adsorbent, c the concentration at equilibrium of the dissolved species in solution, expressed as mmol/L and q the amount at equilibrium of the sites filled with species A adsorbed on the adsorbent, expressed as mmol/g of adsorbent. By defining Θ the fraction of the sites occupied by A, $\Theta = q/(Q_{max})$ where $Q_{max} = (S) + q$ is the maximum adsorption capacity of the adsorbent, we have:

$$\Theta = q\,/\,[(S)+q] \qquad (2.23)$$

Combining (2.18) and (2.19) we obtain:

$$K = \Theta\,/\,[(1-\Theta)\,c]$$

and

$$\Theta = Kc\,/\,[\,1+Kc\,] \qquad (2.24)$$

These equations describe the Langmuir isotherms.
From eq. (2.19) and (2.20) we obtain

$$\frac{1}{q} = \frac{1}{Q_{max}} + \frac{1}{KcQ_{max}}$$

A plot of $1/q$ vs $1/c$ gives a slope of $1/KQ_{max}$ and an intercept of $1/Q_{max}$.

When there are two compounds present in the solution, the equilibrium isotherms are obtained in a similar way. Consider two species, A and B, competing for the same sites of the adsorbent. We have:

$$S + A \rightleftharpoons SA \qquad \text{and} \qquad S + B \rightleftharpoons SB$$

With equilibrium constants:

$$K_A = q_A/c_A(S) \qquad (2.25) \text{ and}$$
$$K_B = q_B/c_B(S) \qquad (2.26)$$

We also have:
$$Q_{max} = (S) + q_A + q_B \qquad (2.27)$$
$$\Theta_A = q_A/[(S) + q_A + q_B] \quad \text{and} \quad \Theta_B = q_B/[(S) + q_A + q_B] \qquad (2.28)$$

Combining eq. (2.25), (2.26) and (2.27) we obtain:

$$\Theta_A = \frac{K_A c_A}{1 + K_A c_A + K_B c_B} \qquad (2.29)$$

and

$$\Theta_B = \frac{K_B c_B}{1 + K_A c_A + K_B c_B} \qquad (2.30)$$

Combining equations 2.27, 2.28, 2.29 and 2.30 we obtain:

$$\frac{1}{q_A} = \frac{1}{Q_{max}} + \frac{1}{K_A c_A Q_{max}} + \frac{K_B c_B}{K_A c_A Q_{max}}$$

and

$$\frac{1}{q_B} = \frac{1}{Q_{max}} + \frac{1}{K_B c_B Q_{max}} + \frac{K_A c_A}{K_B c_B Q_{max}}$$

These equations give the loading of the adsorbent as a function of the solution composition.

By setting $c_A + c_B = C$ and $x_A = c_A/C$ these equations become:

$$\frac{1}{q_A} = \frac{1}{Q_{max}} + \frac{1}{K_A x_A Q_{max}} + \frac{K_B(1-x_A)}{K_A x_A Q_{max}}$$

and

$$\frac{1}{q_B} = \frac{1}{Q_{max}} + \frac{1}{K_B(1-x_A) Q_{max}} + \frac{K_A x_A}{K_B(1-x_A) Q_{max}}$$

The other well known isotherm is that of Freundlich, expressed by the equation 2.31:

$$q = K_f \, c^{1/n} \tag{2.31}$$

or: $\qquad \log q = \log K_f + (1/n) \log c$

K_f is a measure of the adsorption capacity and n a measure of the intensity of the adsorption. If n is between one and ten it indicates favourable adsorption. These parameters can be obtained by plotting on a logarithmic scale q $\underline{vs}$ c.

Figure 2.9 illustrates two equilibrium isotherms of the Langmuir model having different isotherm parameters. The red colored isotherm has higher K value (more favourable equilibrium) than the blue one. As it will be discussed later in this chapter (see figure 2.12), the more favourable isotherm will lead to a steeper breakthrough curve in column operation and therefore to a higher operating capacity. The low concentration part of the isotherms of figure 2.9 indicates the expected minimum leakage when the adsorbent, after the elution step, still contains some residual concentration of adsorbate. The minimum leakage will then be the concentration of the adsorbate in solution in equilibrium with the adsorbent containing the residual amount of adsorbate after elution.

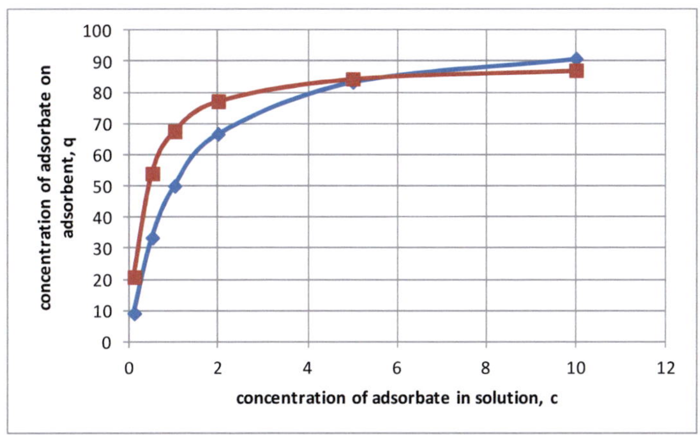

Figure 2.9 Two isotherms, one (red) with higher affinity for the adsorbent than the other (blue).

Adsorption is in general exothermic and therefore the effect of temperature on adsorption is negative. For a given concentration of the adsorbate in solution, an increase in temperature leads to a decrease in the quantity of the adsorbate on the adsorbent. If this effect is significant, then the adsorbent could eventually be eluted by allowing heat in some form to pass through the loaded adsorbent.

The difference between ion exchange and adsorption is that the kinetics of adsorption is slower than the kinetics of simple ion exchange so that the slope of the breakthrough curve in adsorption is smaller than that of ion exchange and in addition, adsorption depends strongly on parameters such as specific flow rate (BV/h), temperature and particle size of the adsorbent. Also, as it may be expected, the adsorption capacity of adsorbents having small pores decreases as the size of the molecules to be adsorbed increases (Adachi 2004). In these cases, the effect of flow rate is even greater, that is, the adsorption of large molecules by adsorbents having small pores is very sensitive to the flow rate. The regeneration in these cases is also affected by the size of the adsorbed compounds. Large molecules are eluted with more difficulty from adsorbents having small pores so that it is possible in these cases that the adsorbents become fouled after a number of cycles.

Ion exchange resins as catalysts

Strong acid cation and strong base anion exchange resins with sulphonic groups in the H^+ form or quaternary ammonium functional groups in the OH^- form respectively, show similar chemical characteristics with inorganic acids and bases. It can be assumed that the counter ions (H^+ and OH^- respectively) of the SAC and SBA exchangers are as reactive as the H^+ and OH^- ions in the homogeneous catalysis and that the reaction mechanisms are similar. Still, there are significant differences

56

between homogeneous and IER catalysis. In IER catalysis, the reactants must diffuse through the resin beads to reach the active sites. Differences in diffusion rates or in the solubilities of the reactants in the resin matrix can result in different selectivites in the case of parallel reactions and into different degrees of conversion of the resin-catalysed reactions with respect to homogeneous catalysis. These, together with process advantages such as the easiness of separation from the reaction medium, the safer storage, handling and disposal contributed into the early use of IER as catalysts in organic synthesis (Albright and Jakovac, 1985).

The main limiting factor of the use of IER as catalysts is their thermal stability. This factor resulted in a much broader use of SAC exchangers, in temperatures up to 150°C, than of the SBA exchange resins which are limited to temperatures up to about 60°C.

By catalytic activity of IER is meant the rate of the catalysed reaction. This depends on the one hand on the type and concentration of the active sites of the IER and, on the other, the accessibility of these sites.

Although the type of the active sites is defined by the chemical structure of the functional groups of the ion exchanger, that is sulphonic groups, quaternary ammonium groups etc., their concentration may vary either by grafting more active sites per monomer unit of the polymeric matrix or by increasing the dry apparent density of the resin, that is the amount of dry solids found in a given resin volume.

The accessibility of the active sites by the reagents depends on the solubility of the reagents in the resin as well as on their diffusivity through the resin matrix.

The solubility of the reactants in the resin, reflecting the interactions between the reactants and the resin matrix, affects the distribution of the reactants and the products between the reaction medium and the resin. As a result, depending on the solvent and the distribution of the reactants between resin and solution, the activity of the resin can become higher or lower compared to the homogeneous catalysis (Pitochelli, 1975). Similar considerations lead to selectivity differences between resin and homogeneous catalysis, when the solubilities of the different reactants in the resin are not the same. In the case of having two reactants, A and B, where two parallel reactions are possible, A reacts with B or with itself, then if the distribution of A and B between resin and solution is not the same for A and for B, for example if reactant B is more concentrated inside the resin, then reaction A with B is favoured in comparison to the reaction A with A while in homogeneous catalysis, both reactions take place.

When the rate of diffusion of the reactants into a gel-type resin or the products out of this resin is faster than the rate of the reaction itself, then the overall rate is controlled by the reaction rate throughout the particles (Helfferich, 1962). In this case, all the catalyst active sites can be used and the catalyst is used efficiently. If, on the contrary, the rate of diffusion of the reactants is slower than the reaction rate, then the controlling step is the diffusion of the reactants in which case only a layer of the resin, near the external surface, is used. The overall rate is controlled by the reaction at or near the resin surface. The catalyst is not used efficiently since active sites found in the centre of the resin beads do not have the chance to see any reactants.

The parameters that influence the diffusivity of the reactants and products are: the cross-linking density of the resin in connection with the size of the reactant and produced molecules; the presence of a reaction mediumthat swells the resin. A low-crosslinked resin that swells strongly shows a low volume capacity (low concentration of active sites) and fast diffusion rates so that the overall kinetics is controlled by the reaction rates throughout the particle. The particle size of the resin does not influence the reaction rates throughout the particle . On the other hand, the particle size influences the rate of the reaction in the case where diffusion through the particle is slow and the reaction rate is controlled by the reaction at or near the resin surface.

The above considerations (diffusivity of the reactants and products, cross-linking density and the presence of a reaction medium that swells the resin) apply also for the organic reactions catalysed by IER. The solvation of the resin matrix along with the size of the reactants and products must first be taken into account. For example, if the IER is poorly solvated in the reaction medium (non-polar media), then the gel structure is collapsed and diffusion rates become very slow. In those cases, only MR resins can be used. In this case, the higher the internal surface area is, the higher the activity of the resin catalyst will be since a higher fraction of the total resin capacity will be found close to the internal surface.

In general, there is not just one parameter that influences the rate of a reaction, but more than one at the same time. Thus, increasing the surface area certainly increases the fraction of sites near the surface but at the same time, since this is due to an increase in the DVB content, it decreases the diffusion of the reactants and products through the resin matrix. Therefore, with a reaction that takes place in the resin matrix (particle controlled), by increasing the surface area the activity of the resin should first decrease until the kinetics become mainly surface controlled, whereby the activity will increase. If the reaction rate in the matrix is slow (reaction at or near the surface of the resin), then increasing the surface area the reaction rate should increase. However, as the DVB level increases, the functionalisation (the total concentration of the active sites) of the polymer matrix decreases, so the reaction rate, after an initial increase, can pass through a maximum to decrease again.

Ion exchange resins as supports for enzyme immobilization

Enzymes are proteins that catalyze biochemical reactions and are used in vitro in many industries including Food. Their high specificity for certain reactions are due to their three-dimensional structural conformation. They can be used in solution but a standard method for using them is to immobilize them on a solid support which has the advantages of multiple reuse while preserving their activity, decreased sensitivity to parameters such as pH and temperature and an easy separation from the reaction medium making possible the development of continuous processes. Immobilization techniques include covalent binding, achieved by a chemical reaction between the aminoacid groups of the enzyme and the functional group of the support, and non-covalent binding which includes hydrophobic, ionic and non-specific interactions.

Ion exchange resins and polymeric adsorbents have been used as supports for enzyme immobilization especially after the development of macroreticular type structure. Resin parameters that affect enzyme immobilization include pore size and internal surface area. Depending on the net charge of the enzyme, cation exchange or anion exchange resins can be used which combined with their pore structure allows the conservation of the steric configuration of the enzyme. Examples of enzyme immobilization in food processing industries include:

Anion exchange resins: glucose isomerase

Polymeric adsorbents: β-amylase, β-galactosidase

Cation exchange resins: lysozyme (recovery)

Glucose isomerise (GI) is a widely used application of immobilized enzymes for the isomerisation of glucose to fructose. Duolite® A568 anion exchange resin from Rohm and Haas Company (now The Dow Chemical Company) having formophenolic matrix has been used industrially in this case.

GI is immobilized on the ion exchange resins by electrostatic adsorption. The glucose syrup, containing at least 93% glucose is adjusted at 40-50 % dry substances (DS)[4], pH 7.5-7.8, 1.5 mM Mg^{2+} and 2 mM HSO_3^- and allowed to pass through the immobilized GI at a temperature about 58°C. The exiting syrup contains 42-45% fructose. The pH is immediately adjusted to 4-5 in order to avoid unwanted base-catalyzed reactions and color formation (Aehle, 2007).

Another way to produce a mixture of glucose and fructose is by hydrolysis of sucrose, called sugar inversion, achieved either by acid catalysis or enzymatic hydrolysis with the enzymeinvertase. Invertase immobilization on anion exchange resins (Ribeiro and Vitolo, 2005) was investigated using Dowex® resins and the hydrolysis of sucrose was evaluated in parallel with both soluble and immobilized enzyme. One of the resin parameters studied

---

[4] Percent dry substances % DS is the (dry mass)/(initial mass) x100.

was the DVB content of the polymer matrix (Tomotani and Vitolo, 2006).

In certain cases, a combination of ionic binding and covalent bonding is more efficient for the residual activity of the immobilized enzyme (Guidini *et al*, 2010). Thus, Aspergillus oryzae β-galactosidase was immobilized on Duolite® A568 of Rohm and Haas Company and the immobilization was studied with ionic binding only and with a covalent bonding using glutaraldehyde. It was found that the residual activity of the enzyme immobilized with ionic interactions only was 51% after 30 cycles while with the enzyme immobilized with glutaraldehyde crosslinking was 90%.

Synergy of ion exchange and membranes

In many of the applications of ion exchange in processing food, a treatment on membranes, usually prior to the IER, becomes more and more frequent, as new membranes are developed with new materials and smaller size pores. The purpose of such treatments is either to purify a solution or to concentrate certain ingredients. By doing this, the IX flowsheet or dimensions of the IER plant may change. Among the membrane types, the ones that risk to affect most the IX treatments are microfiltration (MF), ultrafiltration (UF) and nanofiltration (NF).

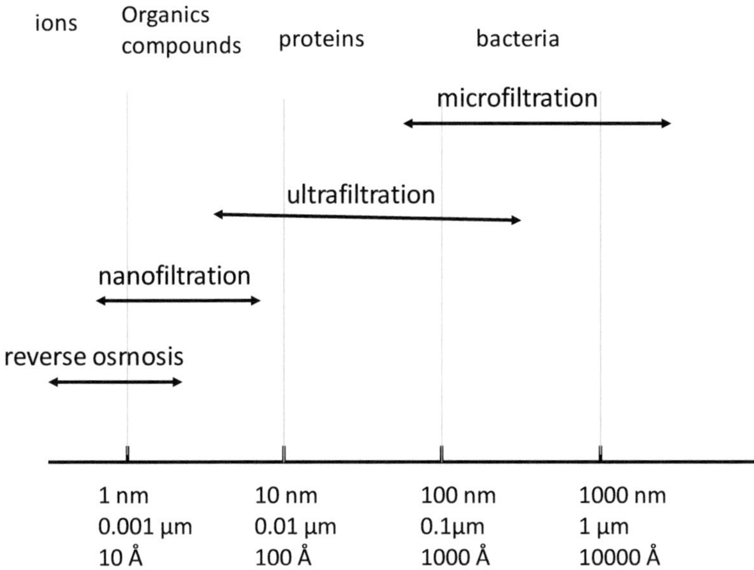

MF membranes have pore sizes range from 0.1 to 1 µm. they can reject particles such as bacteria or suspended solids. They find application in eliminating impurities from the cane sugar clarified juice which then can be decalcified by IX before evaporation.

UF membranes have pore sizes from 0.01 to 0.1 µm, or a molecular weight cut off (MWCO) of 1.000 to 300.000 Da. They can reject proteins. UF is used in milk whey in the recovery of protein concentrates. Other applications include orange juice debittering or deacidification, in separating the pulp from the rest of the juice which is then treated by IER and in the production of white sugar directly from the sugar mills.

NF membranes have pore sizes from 1 ro 10 nm, or a MWCO of 200 to 1.000 Da. They are used partial demineralization by eliminating divalent ions (Ca^{2+}, Mg^{2+}), heavy metals, colour compounds or pesticides. They find applications in milk whey partial demineralization (decalcification) thus allowing a better use of the IER. They also find application in spent regenerant recycling in cane sugar decolorization.

Reverse osmosis (RO) on the other hand, does not really have pores. The water passes through by dissolving into the membrane driven by the high pressure difference between the feed side and the permeate side. Rejection of all ions is high but divalent ions are rejected more than monovalent. Small uncharged organic molecules such as methanol, ethanol etc have low rejection efficiency and pass through..

Ion exchange processes

Ion exchange between resin and solution can be achieved by different techniques:
- batch operation
- column operation
  - single fixed or fluidized bed
  - two or more columns in series
- continuous systems

In a batch operation, the entire solution to be treated is mixed with an IER under agitation until after a certain time equilibrium is reached. The extent to which the exchange of ions will take place depends on the selectivity coefficients or the separation factors for the given ions. Consider for example the exchange reaction (2.1) where initially a SAC resin is in the H^+ form and the solution contains only NaCl.

$$\mathcal{R}\text{-}H + Na^+Cl^- \rightleftharpoons \mathcal{R}\text{-}Na + H^+Cl^- \qquad (2.1)$$

After mixing certain volumes of resin and solution and waiting for equilibrium, the resin will be found partially in the H^+ form and partially in the Na^+ form while the solution will contain a mixture of NaCl and HCl. For a given solution to resin volume ratio, the exact resin and solution composition will depend on the value of $K_{Na/H}$. The higher the value of $K_{Na/H}$ is, the more Na^+ ions will go onto the resin and the less will remain in solution. With only one batch contact however, the Na^+ ions will never go 100% onto the resin, some Na^+ concentration will always remain in solution. If we want to remove a high fraction of the Na^+ from the solution, a second batch contact has to be carried out with the same solution after equilibrium from the first contact and with a fresh resin in the H^+ form. After this second contact, the Na^+ ions remaining in solution will be less than after the first contact. These batch contacts can be continued for a number of additional contactors containing fresh resin, until the desired level of Na^+ in solution is reached. The situation will be different if we continue to pass fresh solution through the above batch containers, but leaving the resin as it was from the previous solution. As fresh NaCl solution enters the first compartment, equilibrium will take place with the already partially loaded resin, and so on with the resin in the rest of the compartments.

After each new equilibrium, the resin in each compartment will contain more Na^+ ions and less H^+ ions. If this continues many times, a point will be reached that the resin in the first contactor will contain practically only Na^+ ions. We see in other words that if we continuously treat a fresh solution volume with a certain resin quantity, we can almost fully utilize the full resin capacity.

In column operations, a solution containing ions to be removed flows through a column containing the resin in a certain ionic form.In a single fixed bed column operation the solution flows continuously through a fixed resin bed, usually down flow. One can visualize the resin being a series of layers, or plates, with which the solution comes successively in contact, in a similar way as with a stirred tank in a batch operation. Depending on the kinetics of the ion exchange, after a certain contact time, an approximate equilibrium is reached in each layer. After contacting the first IER bed layer, the solution with the remaining ions from the first contact goes on further to the second layer, where it again moves towards equilibrium, resulting in an even lower concentration of the ions remaining in solution. As the solution progresses down through the resin in the column, the successive contacts with layers of the resin further reduce ion concentration in solution until it attains a constant value. As fresh solution comes into the resin column, similar contacts take place with partially loaded resin and more of the incoming ions are removed. After a certain volume of feed solution has passed through the column, the upper part of the resin bed will become exhausted (loaded) and its composition will remain constant. The resin zone in which the concentration of the ion to be removed in the solution drops from the feed value to the final value is called ion exchange zone (IEZ) or reaction zone or front etc and is illustrated in figure 2.10.

Consider a binary ion exchange. If the equilibrium is favourable, and depending on the value of the separation factor, a reasonably sharp IEZ will be formed. As the solution moves down the resin bed in the column, a steady state is reached where the IEZ moves down without changing length (Fig.2.10a). This is called a "self-sharpening front". When a pre-determined ion concentration is found at the column outlet, the loading cycle usually stops (Fig. 2.10b). This is termed the end-point, or breakthrough point, or operating capacity point.

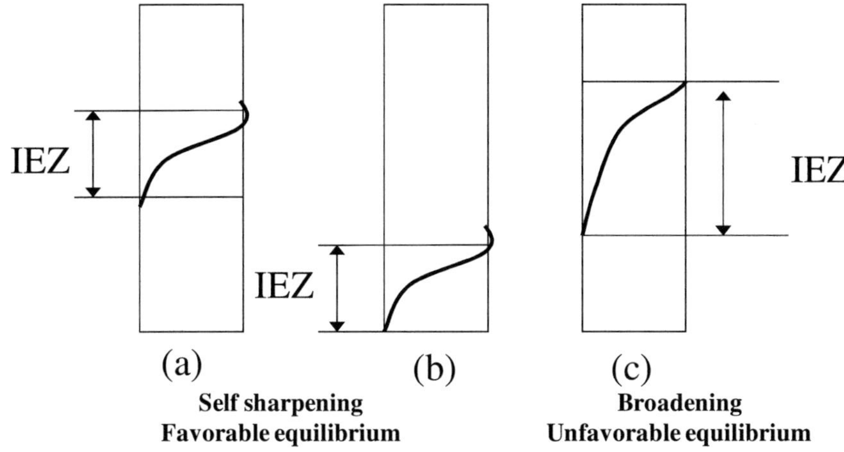

Figure 2.10 Ion exchange zone

If the equilibrium is unfavourable, the ion concentration in solution diffuses as it goes down the resin bed and it never reaches a steady state (Fig. 2.10c). This is called a "broadening front".
It should be noted here that the formation of these fronts assumes a uniform resin bed free from imperfections,

preferential channelling, foreign matter in the resin bed, air pockets etc, so that the solution flows through like a piston.

Loading cycle

Another way of presenting column operation is by plotting the concentration of the ions exiting the column against the volume passed through the column, or versus time. Provided that the liquid flow rate is constant, these two ways are equivalent. In order to normalize the volume, Bed Volumes (BV) are used where:

$$BV = (\text{Volume of solution}) / (\text{Volume of resin})$$

Similarly, for flow rate, bed volumes per hour, BV/h, is used. BV/h is called specific flow rate, as opposed to the volumetric flow rate, volume of solution per unit of time. The contact time of solution with resin is inversely proportional to the specific flow rate. Recalling that the linear velocity of the liquid is equal to the volumetric flow rate divided by the column cross sectional area, one can easily derive that the linear velocity is equal to the specific flow rate multiplied by the resin bed height.

In multicomponent systems it is generally assumed that the exchange of each component does not interfere with the exchange of the other components. Consider the case where the feed solution contains NaCl and KCl and the resin initially is 100% in the H^+ form. The selectivity coefficient $K_{K/H}$ is higher than $K_{Na/H}$. Initially, both Na^+ and K^+ will be exchanged for H^+ on the resin and H^+ will be displaced into the solution. From the selectivity coefficient we see that the resin prefers K^+ from Na^+. As a consequence, the K^+ will displace some Na^+ with the result that Na^+ will be predominant in solution. Two IEZ will be formed, one with K^+ / Na^+ followed by one with Na^+ / H^+. The resin above the first IEZ (that of K^+ / Na^+) will be partially in the Na^+ and partially in the K^+ form with the K^+ the predominant ion. Below the Na^+ / H^+ zone, the resin will be in the H^+ form. When the Na^+/ H^+ IEZ reaches the bottom of the column, Na^+ will start leaking and will eventually reach a maximum ("chromatographic peaking") due to the displacement of Na^+ by the K^+ ions, before dropping to the feed concentration levels (figure 2.11). At the point where the K^+ / Na^+ zone reaches the bottom of the column, K^+ ions will start to leak. At saturation, the resin will be in a mixed K^+ and Na^+ form, the exact composition of which depending on the solution composition and the value of the separation factors.

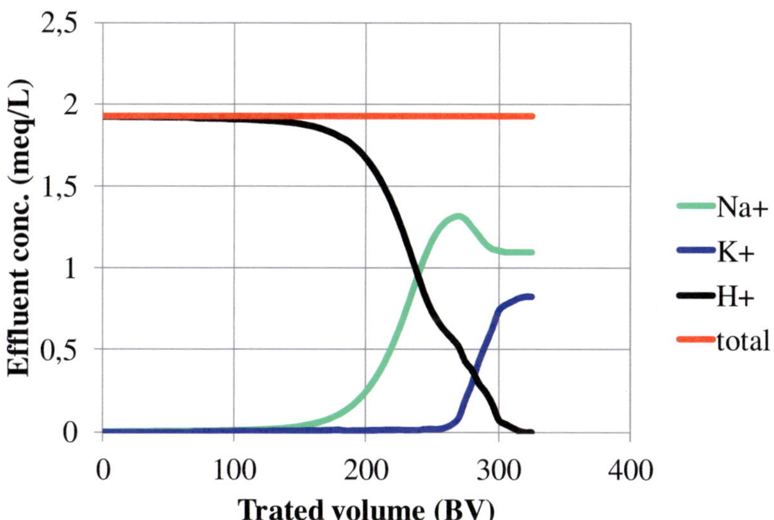

Figure 2.11 Breakthrough curves for the reaction
R- H + Na^+Cl + K^+Cl $\rightleftharpoons$ R- Na + R-K + H^+ Cl^-

The IEZ of a self-sharpening front, illustrated in fig. 2.10a and b, corresponds to the distance from the end-point to saturation in a breakthrough curve, for example from 170 BV to 250 BV for the Na^+ leakage curve in figure 2.11.

In the hypothetical case where the selectivity coefficient is very high, the liquid flows through the resin bed like a piston and the ion exchange takes place instantaneously, the length of the IEZ will be extremely short and the IEZ in figure 2.10 becomes extremely sharp. In practice however, the imperfections of the resin bed caused by preferential channelling, axial dispersion of the ions or imperfections in the resin beads structure result in an increase in the length of the IEZ. In addition the selectivity coefficient and the kinetics of ion exchange cause further IEZ length increase. Factors such as flow rate and resin particle size that affect the extent of the equilibrium reached in each plate will affect the length of the IEZ. This is illustrated in figure 2.12 which shows two breakthrough curves, one where the kinetics of exchange is slow and/or the selectivity coefficient is low (curve I) and the other where kinetics are fast and/or the selectivity coefficient is high (curve II).

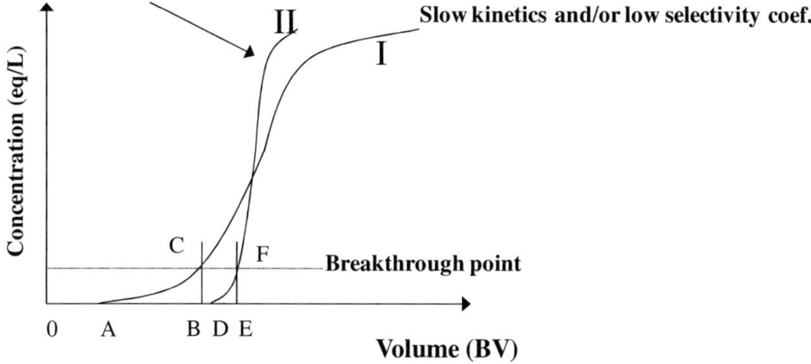

Figure 2.12 Breakthrough curves with slow (I) and fast (II) kinetics or low (I) and high (II) selectivity coefficients

As shown in figure 2.12, when the leakage reaches a certain value, the breakthrough point, then the loading cycle stops and this determines the operating capacity of the resin.
The operating capacity (op cap) can be calculated using the expression:
Op cap = (concentration of the fixed ions)*(treated BV)

For curve I of fig 2.12 the operating capacity is:

op cap = 0B * (feed conc.) – (surface of ABC)

Similarly, the operating capacity for curve II is:

op cap = 0E * (feed conc.) – (surface of DEF)

That is, curve II results in a higher operating capacity in comparison to curve I. Thus, a resin showing the characteristics

71

of curve II will produce a longer service cycle and lower leakage. Often it is advantageous to have such a resin or to vary the operating conditions such that breakthrough curves similar to that of curve II are obtained.

The particle size of the resin has an important impact on the kinetics of ion exchange and therefore on the operating capacity. Small size beads present a large surface area through which the ions can enter into the beads and diffuse through the particles. In addition, the Nernst film is thinner with small beads and the distance that the ions have to cover inside the resin particle to reach an exchange site is shorter. Consequently, resins with smaller particle size show faster kinetics and have higher operating capacities.

In a similar way, the breakthrough curve is also affected by the flow rate. In particle diffusion controlled kinetics, by increasing the volumetric flow rate of the solution through a given volume of resin, in other words by increasing the specific flow rate (BV/h), the incoming ions remain in contact within a layer of the resin bed for a shorter period of time, and thus move to the next stage prior to equilibrium being reached. The IEZ therefore becomes longer.

As suggested in figure 2.10 (b), the higher the resin bed height in relation to the length of the IEZ, the larger the fraction of the ionic sites which will be used. Naturally, this implies that more resin will be needed to treat a given volumetric flow rate. What it should be avoided is to have a resin bed height shorter than the IEZ length as this will result in an unacceptably short cycle time. In cases where the IEZ length is high, two or even three columns in series can be used.

Ion exchange is a cyclic process. Because the ion exchange reaction is reversible, when the loading cycle ends, a regeneration, or elution step follows where the resin is stripped of its loaded ions and converted back to its initial ionic form. For example, if the loading cycle consists in removing Cl^- from a solution with a SBA resin in the OH^- form, eq (2.32), the regeneration consists in converting the resin back to the OH^- form using a regenerant solution such as NaOH (2.33).

$$\mathcal{R}\text{- OH} + Cl^- \rightleftharpoons \mathcal{R}\text{- Cl} + OH^- \qquad (2.32)$$
$$\mathcal{R}\text{-Cl} + Na^+OH^- \rightleftharpoons \mathcal{R}\text{-OH} + Na^+Cl^- \qquad (2.33)$$

In an ion exchange reaction, if loading involves a favourable equilibrium then the regeneration (elution), which is the same reaction but in the opposite direction, should be an unfavourable equilibrium. For the above exchange, eq. 2.24, the separation factor for a gel type SBA resin, $\alpha_{Cl/OH}$, is about 15. This means that for the reverse reaction, eq. 2.25, the separation factor $\alpha_{OH/Cl}$ is $1/15 = 0.067$, and therefore the reaction is unfavourable. The elution profiles for the two cases are shown in figure 2 13.

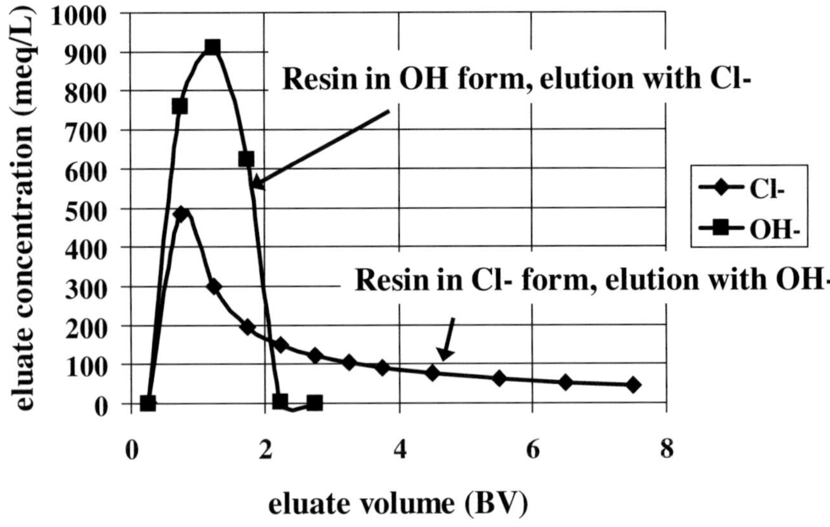

Figure 2.13 Elution profiles, flow rate 2 BV/h, eluent 1N NaOH or 1N NaCl, gel type SBA resin

As a result, the regeneration conditions, such as regenerant concentration and temperature, are chosen so that regeneration becomes as efficient as possible. In fact the quantity of regenerant employed is often a question of economics. The more regenerant used, the higher fraction of the resin will be converted to the initial ionic form. However, as it is seen from figure 2.13 above, most of the fixed ions are eluted during the first part of the regeneration while a large excess of regenerant is needed to remove the fixed ions down to very low level. Therefore, the regenerant level to be used is dictated by the cost and by the quality of the product solution during loading, in other words by the leakage tolerated.

74

One special condition which considerably affects the regeneration efficiency is the direction of the regenerant flow with respect to the feed solution during the loading step. As it can be seen in figure 2.10(b), with a down flow loading at the termination of the loading cycle the bottom part of the resin bed is only partially loaded while the upper part is saturated. When the regenerant flows down flow (co-flow or co-current regeneration, in the same direction as the feed flow), when the spent regenerant passes through the lower part of the resin bed, it saturates the partially loaded resin before it exits the column. Thus, the regenerant has not been fully utilized. Furthermore, in order to regenerate the lower part of the resin bed as completely as possible, a large excess of regenerant is necessary. If on the other hand the regenerant flows upward (counter-flow or counter-current regeneration, in the opposite direction to the feed flow), the lower part of the resin bed remains well regenerated whilst the upper part of the bed is only partially regenerated, which however does not affect significantly the effluent leakage during the following loading cycle. The subsequent leakage in the following down flow loading cycle will be much lower in counter-flow regenerated beds than in co-flow regenerated beds.

In general, the regenerationof ion exchange resins consists of :
- a backwash
- the regenerant injection
- a slow (or displacement) rinse
- a fast rinse

The backwash step is designed to remove any foreign particles or resin fragments and to thus clean and decompact the resin bed. This ensures that the in-service resin bed pressure drop is kept to the minimum. For an efficient backwash, the free space above the resin bed should be approximately 60% or more and this expansion should be maintained for 15-20 minutes or until the resin bed is clean.

The regenerant injection time should be sufficient for an efficient use of the regenerant. In the case of small ions with fast kinetics, this is in the range of 30-40 minutes for a standard particle size resin.

The slow rinse is designed to displace the regenerant still remaining in the resin bed and is carried out at the same flow rate as the regenerant flow rate.

The fast rinse is designed to remove the last traces of regenerant and is normally carried out at a flow rate of at least 80% of the normal service flow rate.

Typical regeneration sequences of the styrene-DVB adsorbents in the case where chemicals are used as regenerants are the following:

- sweetening off: demineralized water pushes the juice out of the adsorbent until the effluent is less than 1°Bx. At a specific flow rate of 4 BV/h it should take about 2 BV of water or 30 minutes.
- Backwash (40 min)
- Chemical injection: 2 BV of a 2 % NaOH solution at 2 BV/h. (1 h)
- Slow rinse: demineralized water pushes the regenerant out of the adsorbent at the same velocity as the regeneration. This ensures the efficient end of the regeneration. 2 BV are sufficient (60 min)
- Acid injection: 2 BV of 1% HCl at 2 BV/h (1 h)
- Slow rinse, 1 BV at 2 BV/h

- Sweetening on: juice passes through the cation, anion and adsorbent resins until at the exit of the adsorbent the juice concentration reaches >90% of the feed concentration.

One reason for the acid wash is to lower the pH of the adsorbent after the NaOH injection. In principle, adsorbents do not have functional groups and therefore, the slow rinse after NaOH injection should be sufficient. However, it may be that some adsorbed compounds containing weak acid groups remain on the adsorbent and these give an alkaline pH, similar to the fouled WBA resins. The other reason is to elute any acid-soluble compounds remaining on the adsorbent after the NaOH treatment.

The start-up procedure of a new styrene-DVB adsorbent consists of the following steps:

Load the resin into the column:

- Fill the column half-way with water
- charge 1/3 of the resin into the column
- open the upflow water valve and allow water to rise at a velocity of about 0.7 m/h to 2 m/h depending on the adsorbent.
- Keep the upward flow for 15 minutes
- Shut off the water valve and allow the resin to settle. Drain the water to about 1 meter above the resin.
- Repeat this operation with the rest of the resin.

Start-up :

- Wash the resin with 10 BV of water to eliminate the salts that the adsorbent contains as a bacteriostat. The quantity of the water (10 BV) is an indication, it also depends on the application (for ultrapure products for example, more water may be necessary to reduce the salt effluent concentration to very low -ppb or lower- levels).

- Start with the regeneration cycle

Phenolic adsorbents require special attention in using them in columns. They have a granular rather than spherical form that most ion exchange and adsorbents have and they can easily agglomerate to form big chunks difficult to disperse. For that reason, they need frequent air scouring to disperse these agglomerates. As an indication, phenolic type adsorbents need the following regeneration sequences:

- sweetening off: demineralized water pushes the juice out of the adsorbent until the effluent is less than $1°$ Bx. At a specific flow rate of 4 BV/h it should take about 2 BV of water or 30 minutes.
- Chemical injection:2 BV of a 2 % NaOH solution at 2 BV/h. (1 h)
- Air agitation
- Rinse at 4 BV/h for one hour.
- Acid injection: 4 BV of 0.5 % HCl at 4 BV/h (1 h)
- Air agitation
- Rinse at 4 BV/h for 30 minutes
- Air agitation
- backwash
- Sweetening on: juice passes through the cation, anion and adsorbent resins until at the exit of the adsorbent the

juice concentration reaches >90% of the feed concentration.

The acid wash here is to neutralize the OH groups of the phenolic adsorbent which after the NaOH injection they are found in the Na^+ form. The indicated level of HCl wash, 4 BV of 0.5% HCl or 20 g HCl/L_R, is about the acid quantity necessary to neutralize the phenolic OH groups. If the adsorbent becomes fouled with time of use, then this level should be increased accordingly.

Three columns in series (merry-go-round)

From figure 2.10 it is seen that if the loading cycle stops at a point where the resin is near saturation, then the operating capacity reaches a maximum corresponding to the equilibrium value of the feed solution concentration. However, by the time saturation occurs, the leakage of the compounds to be removed is very high and probably unacceptable by the process parameters. If the cycle is stopped at the breakthrough point and if the IEZ is very long, then only a small fraction of the total resin capacity is utilized during the loading cycle. In order to have both, good resin utilization and low leakage, a second column, with the same resin volume as the first column, can be added in series. The ions leaking from the first column will be taken up by the second column which acts as a polisher, to give the lowest on-line leakage condition. When breakthrough is evident at the outlet of the second column, the cycle stops for a short period of time and the first column goes to regeneration, the second column becomes the primary column and a third, freshly regenerated, column goes to the polishing position. The loading cycle then continues. This system is called merry-go-round system. In a well designed merry-go-round system, when the tail column breaks through, the head column is close to saturation.

The advantage of this system is

- in a well designed merry-go-round system, the resin inventory can be less than two single parallel columns.
- Since the head column that goes to elution is loaded close to saturation, the eluate contains a higher concentration of the eluted ions. This results in a more efficient usage of eluent.

Continuous systems

The three column merry-go-round system is the simplest continuous system. Suppose we have, instead of two columns in series, six columns in series where when the last column breaks through, the first one is close to saturation while from the second to fifth, the columns are progressively less and less loaded. When the last column breaks through, the first goes to elution, the second column becomes head column and so on and in the end, a freshly eluted column is added. This configuration necessitates that the time of elution is at least the same or less than the time the last loading column goes to breakthrough. If this is not the case, then more columns need to be added to the elution section to enable fresh eluted columns to be available when the loading column breaks through. For example, if the elution time is three times that of the time to loading breakthrough, then three columns should be in the elution section. In this case, the eluent flows in counter-current direction of the columns that come to elution.

These principles are illustrated in figures 2.14 and 2.15. In figure 2.14, the absorption section contains six columns. The pregnant feed solution flows from column A1 to A6 and from column A6 it goes to the effluents. The shades of the columns in the figure indicate the degree of exhaustion of the resin: the darker the colour, the more exhausted the resin is.

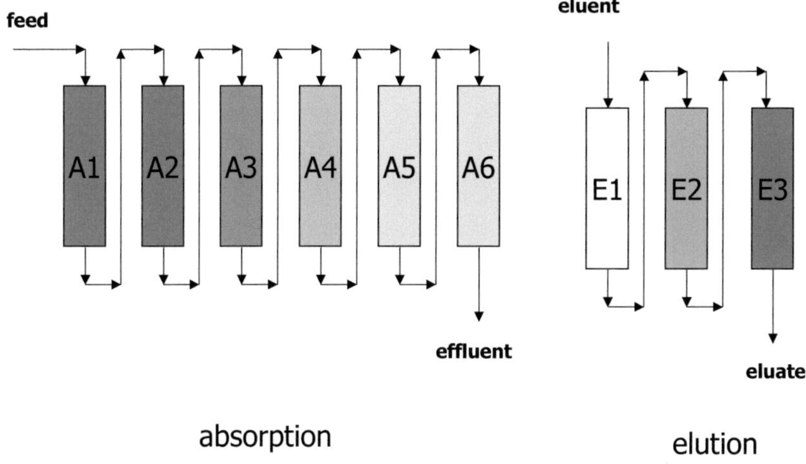

Figure 2.14 Continuous system, principle

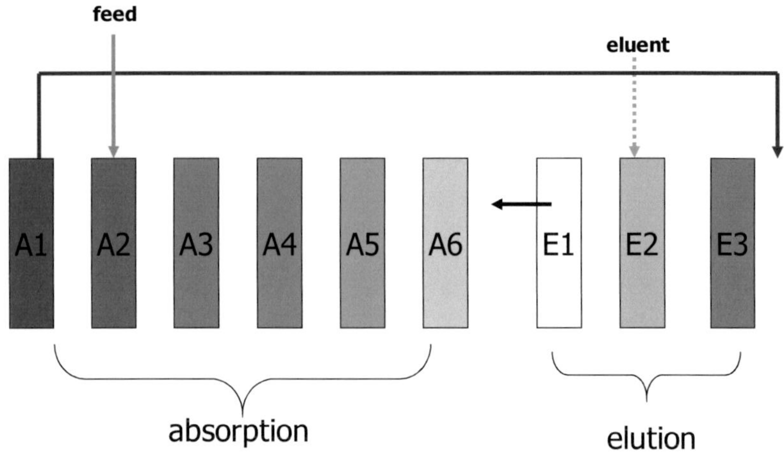

Figure 2.15 Continuous system, resin transfer

The elution (regeneration) column contains three columns and fresh eluent flows from column E1 to E3. From column E3 it goes to the concentrated eluate tank. Every a period of time that corresponds to the time it takes for breakthrough from column A6, say 4 hours, the feed solution flow and the eluent flow stop, column A1 is removed from the circuit and the feed solution is connected to column A2. Column E1 is rinsed and is connected to the exit of column A6 while column A1 takes the place of column E3 (fig.2.15) after it has been rinsed before to take feed solution out.

There exist different continuous ion exchange (CIX) designs. Among the first developed continuous systems is the Higgins Loop® of Severn Trent Services (figure 2.16). The left part of the loop contains the adsorption (loading) part and the regeneration part, separated by valves. In the right part there is a

82

pump that at intervals provides pulses that move the resin around the loop. In the right part there is also the backwash step. The resin and the liquids are moving in the opposite direction, as shown in the figure.

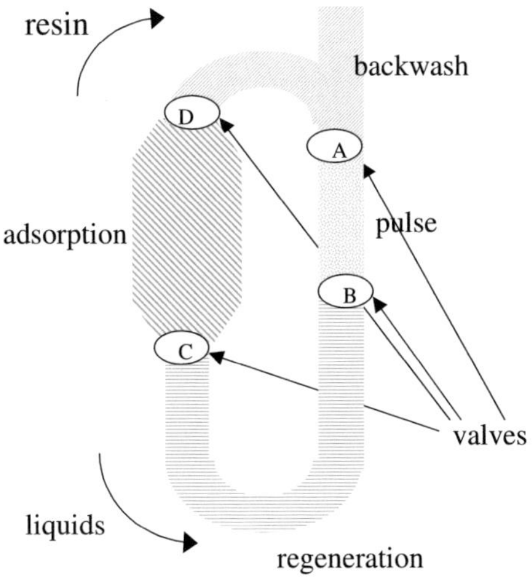

Figure 2.16 The Higgins Loop®

Other continuous systems consist of an arrangement of a number of small cells or columns divided in zones where the process steps, loading, rinse, regeneration and rinse are taking place. In one of them, ISEP® system of Calgon Carbon Corporation, there is a central cylindrical distributor containing 10, 20 or 30 ports. Part of the distributor is fixed while the other part rotates at a specified rate. The stationary part receives the incoming fluid streams and directs them to the appropriate ports of the rotating part and feeds the cells containing the resin or the adsorbent. The out-going streams go out through the ports and the distributor. The cells (or columns) are arranged in a loop divided into four zones, the exhaustion, rinse, regeneration and rinse. Figure 2.17 illustrates an ISEP system adapted for thin juice decalcification usin the NRS process. In this figure, the steps involved are: softening, sweetening off, regeneration, sweetening on, softening.

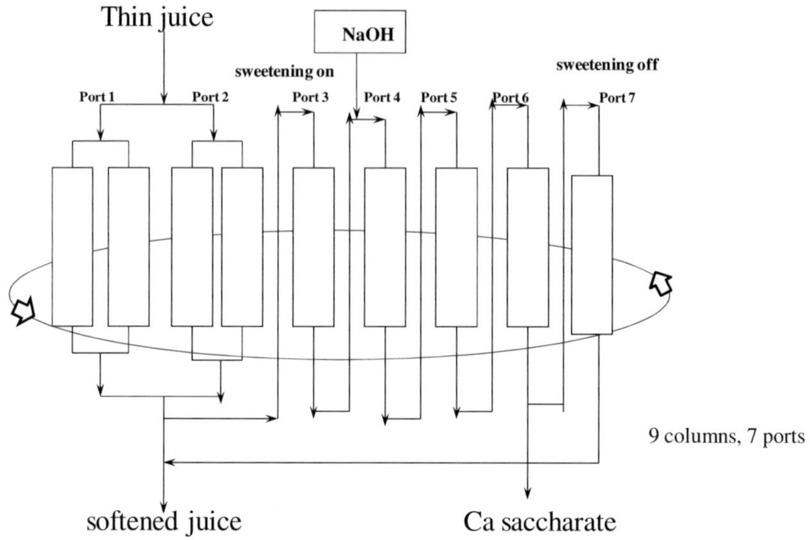

Figure 2.17   ISEP® System for thin juice decalcification using the NRS process.

In the Applexion® ContinuousIon Exchange process (of Novasep) instead of moving the resin or the columns, only the inlets and outlets are shifting. The columns are multi-cells equipped with multi-port valves. The number of cells in each zone varies according to the application.

Frontal analysis and elution chromatography

In what follows in this section about chromatographic separations refers to large scale, industrial, chromatography using ion exchange resins of particle size 200-400 μ or even standard size some times, and columns containing many cubic meters of resin, as opposed to laboratory scale, preparative chromatography.

The column operation discussed above is in fact the so called frontal analysis chromatography, where the sample is fed continuously into the resin bed. The solutes come out in the effluent, first the less strongly fixed compound, then a mixture of the first plus the second less strongly fixed compound and so on. This is the most frequently used technique in ion exchange, as discussed above, even though is the least frequently used technique in chromatographic separations, where displacement or elution techniques are used almost exclusively.

In elution chromatography, a sample containing the solutes to separate is injected on top of the column containing the stationary phase, then an eluent (the mobile phase) enters the column and flows through. The stationary phase may be an ion exchange resin or a support upon which the stationary phase is chemically bonded. The elution can be isocratic, where the eluent has a constant composition, or gradient, where the eluent varies and increases in eluent strength. The solutes are distributed between the eluent and the resin until an equilibrium is reached. As the eluent keeps flowing, the concentration profiles of the solutes in the two phases are also moving. The stronger fixed solute on the resin moves slower while the less strongly fixed solute moves faster down the column.

In a "normal-phase" chromatography, the stationary phase is polar and the more polar solutes are retained longer and come out last. The more polar eluent have a higher elution strength. In "reverse-phase" chromatography, non-polar molecules are grafted on the stationary phase so that it becomes very non-polar. Here, the non-polar solutes are retained longer and come out last, that is the opposite of the normal-phase. The less polar eluents have higher elution strength.

Depending on the interaction mechanism between stationary phase and solutes, we have adsorption, hydrophobic interaction, ion exchange, ion exclusion, ligand exchange or size exclusion chromatography. Among the large scale industrial ion exchange applications are: ion exchange chromatography, for example rare earths separation, ion exclusion chromatography, for example molasses desugarization or glycerine purification, ligand exchange chromatography, for example glucose-fructose separation, and size exclusion chromatography, for example glucose enrichment, where glucose is separated from glucose oligomers.

Consider the case of a chromatogram of two components A and B of a solution where the two components overlap significantly. In order to better separate the two components, either we have to move apart the two peaks, or to make each peak narrower. By increasing the resin bed height (in other words, increase the number of theoretical plates N), the two peaks are moved apart but at the same time the peaks become broader. This approach has the problems that the retention time increases (it takes long time for the peak to come out of the column) and the peaks are broadened, thus resulting in a more diluted solution. This is illustrated in figures 2.16 and 2.17 which illustrate the same two compounds A and B with the same resin but where the bed depth has tripled (from N=10 to N=30).

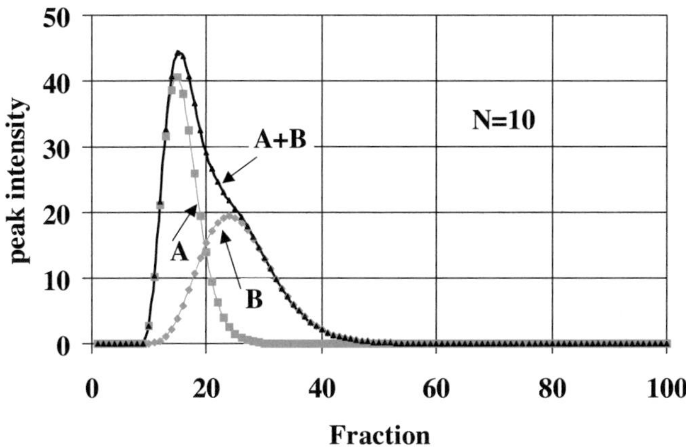

Figure 2.18 The effect of bed depth on separation: N=10

Another way to improve separation is by decreasing the theoretical plate height, thus having more theoretical plates in the same resin bed depth. This can be achieved with operating conditions such as flow rate but also with appropriate resin parameters: these parameters are particle size and crosslinking density (% DVB). This explains the fact that all of the commercial products are resins varying in DVB and particle size.

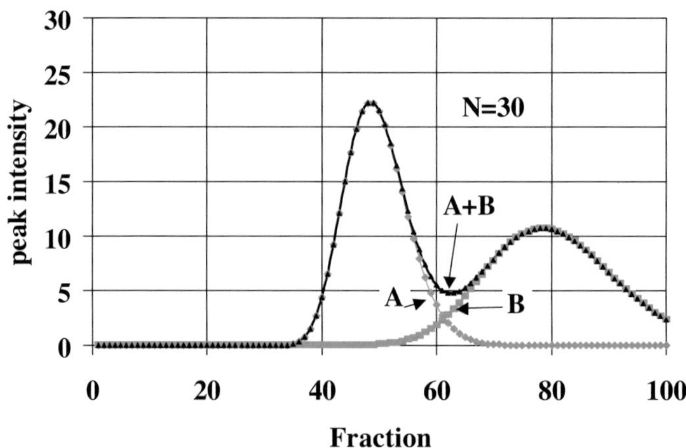

Figure 2.19 The effect of bed depth on separation: N=30

Generally speaking, the lower the DVB is, the better is the separation due to better diffusion of the molecules inside the resin beads. Also, the smaller the size, the better is the separation. The limiting factor is pressure drop, which increases as particle size decreases but also as the DVB decreases and the resin becomes softer. Another limiting factor is the oxidative stability of the resin which results in decrosslinking of the resin by breaking the bonds of the polymer chains. The lower the DVB is, the sooner the effect of resin oxidation (pressure drop increase) will become apparent. The oxidative stability of the chromatographic resins becomes a problem in presence of reducing sugars and of dissolved air (oxygen) in the feed solution.

It can be concluded that it is possible to obtain comparable separations with a higher DVB, smaller resin and a lower DVB larger particle size resin. The final resin choice can be influenced by the dilution of the feed solution during elution and the pressure drop.

In ion exchange chromatography, the solutes to be separated, for example B and C, are ions, fixed on a suitable ion exchange resin initially in form A. B and C have higher affinity for the resin than A. A sample containing the solutes to separate is injected on top of the column where B and C are fixed on the resin, the ion with higher affinity, let's say B, being predominately above the ion with lower affinity, C. Then an eluent enters the column and flows through. The solutes are distributed between the eluent and the resin until an equilibrium is reached. If the eluent is the same as the initial form of the resin, A, then because A has lower affinity for the resin than B and C, the eluent will form a broad front overrunning the two ions B and C which come out the ion C first, followed by ion B (elution chromatography). If the eluent is a different ion, say D, with higher affinity for the resin than B and C, then in the eluate, C will come out first, followed by B, followed by D (displacement chromatography).

Batch chromatography is a simple and economical way of separating compounds. It is a scale-up of an analytical laboratory column for chromatographic separation. However, the peak separation and purity cannot be of very high quality. Continuous systems have been developed that offer a better separation with sharper peaks and lower resin inventories. One system that has been well established is the Simulated Moving Bed (SMB) system, best suited for the separation of two components of a solution.

The idea behind SMB is that eluent moves through the column while the resin bed "moves" in the opposite direction at a speed intermediate between the speeds with which move the two components of the solution. In this way, the two components move in opposite direction, the faster moving along the eluent while the slower moving along the resin bed direction (fig. 2.20). If the feed is added continuously, the two components separate continuously, one moving in one direction and the other in the other direction.

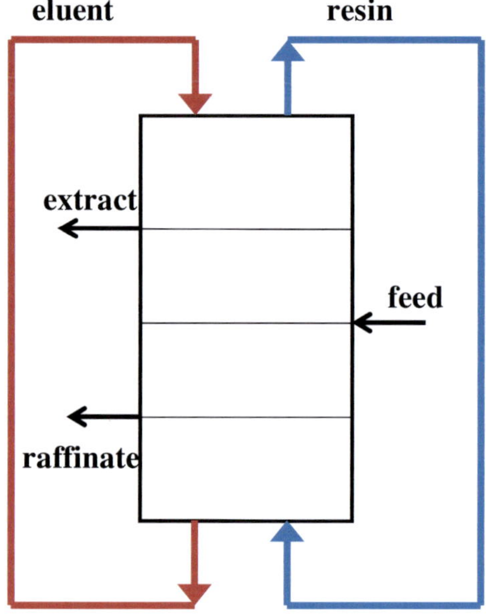

Figure 2.20   Principle of SMB

In reality, it is not the resin bed that moves. The counter-current movement of eluent and resin bed is simulated by shifting periodically the positions at which eluent and feed enter and products are recovered, as illustrated in figure 2.20.

The column consists of a number of compartments, the injection and the eluent enter at given ports and the two solutes exit at two different ports. The port where the slowest moving component exits is called extract while the port of the faster moving component is called raffinate. Periodically, the feed and eluent ports as well as the raffinate and extract ports are shifted towards the direction of the eluent flow. The products are removed by bleeding off a calculated flow at a specified exit port. This is illustrated in figure 2.21

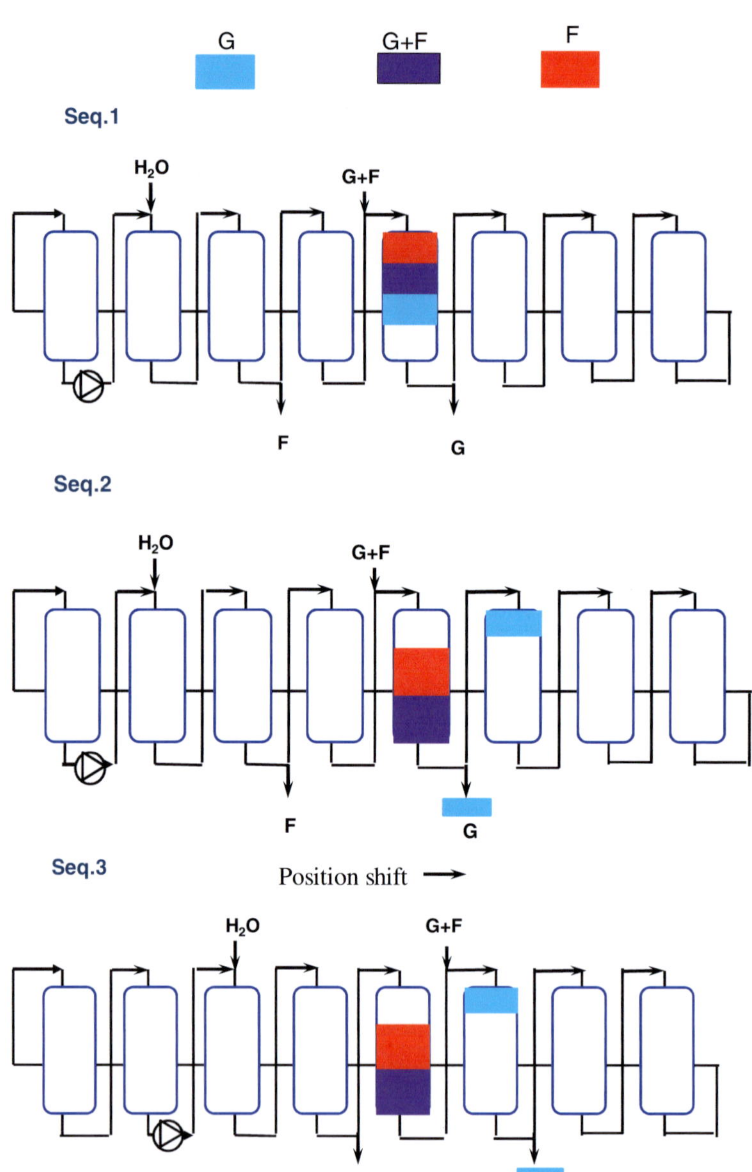
G
G+F
F
Seq.1
H₂O
G+F
F
G
Seq.2
H₂O
G+F
F
G
Seq.3
Position shift
H₂O
G+F
F
G

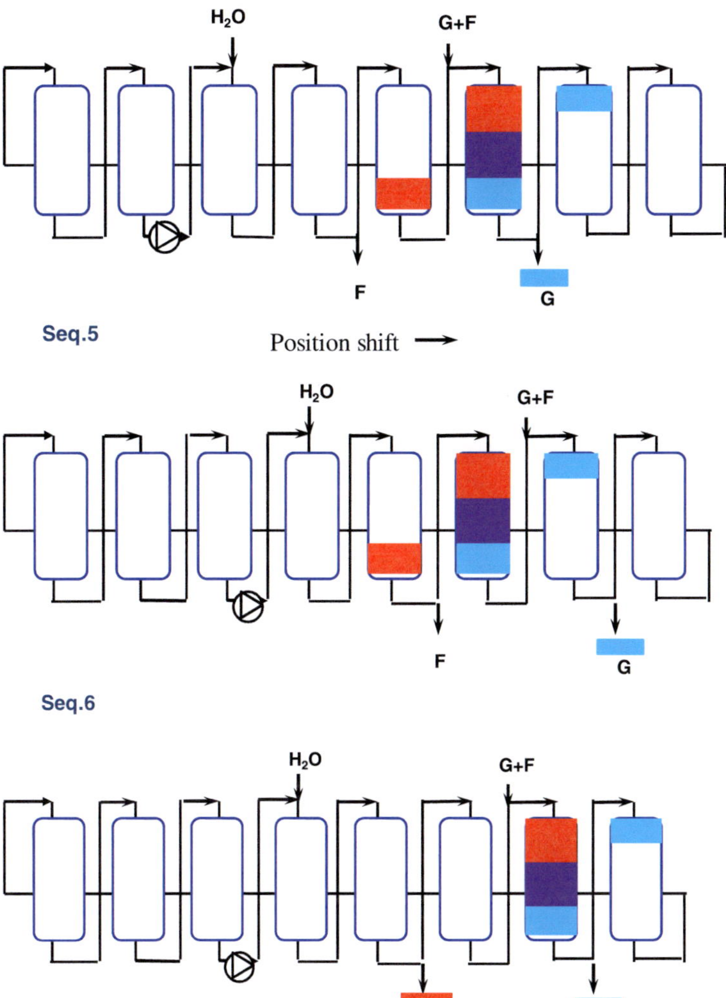

Fig 2.21 SMB illustration

The arrangement of the columns in continuous chromatographic separation systems is similar to the continuous systems mentioned in page 80. The chromatographic separation system of Calgon Carbon Corporationiscalled CSEP® system and the chromatographic separation system of Novasep is mainly the Applexion®SSMB (sequential simulated moving bed). Nippon Rensui Cohas developed a system called ISMB® (Improved SMB) for the separation of two components and the NMCI® system (New MCI system) to separate three or more components of a solution. For separating three components, Novasep has developed, through collaboration between FinnSugar of the Dupont-Danisco group and Applexion of the Novasep group, the NS2P/FAST technology (**N**ew **S**equential **2**-**P**rofile/ **F**innSugar **A**pplexion **S**eparation **T**echnology) while Amalgamated Research Inc (ARi) has developedthe Coupled Loop® process.

The Applexion® SSMB system of Novasep is the second generation SMB continuous chromatography process. The principle of this system consists in simulating a resin flow counter-current to the liquid flow so that the propagation velocity gradient of raffinate and extract is amplified and the separation of the different fractions is improved. The chromatographic cells are operated in series and in closed loop. Feed and eluent injections, as well as extract and raffinate recoveries are sequential and discontinued. The resin bed is fixed, thus the counter-current is simulated moving back periodically both injection and recovery points. After several columns rotation cycles, when the separation profile of the components is stabilized, it is possible to differentiate four elution zones.

94

The Applexion® SSMB chromatography enables purity in the extract and recovery yield to be improved compared to traditional systems (usually 95/95 compared to 90/90, and it is even possible up to 99/99). The process works with only 4 to 8 compartments (fig. 2.22) and works with less installed resin. It is possible to use a finer resin in safe conditions.

Figure 2.22 Applexion® SSMB column arrangement (Courtesy of Novasep)

3. The Sugar industry

Sugar refers to sucrose and comes primarily from sugar beets and sugar cane. IER have been used in sugar processing since more than 70 years. Today IER represent an established unit operation in sucrose production which helps decreasing the production costs and increasing the sugar recovery, thus contributing in improving the profits of the sugar manufacturer (Lancrenon and Hervé, 1988). The increasing use of IER in sugar processing comes together with improved ion exchange products, improved engineering systems and better plant control including wastes disposal when ion exchange process are employed.

The tasks that can be achieved by the use of IER include:
- prevention of scaling in evaporators
- increase the yield of sucrose (S) by removing the non-sugars (NS) or by exchanging melassigenic for less melassigenic cations
- recovery of valuable by-products by chromatographic separation
- inversion of sucrose to form invert sugar
- improve the quality of sugar by elimination of colour

production of high quality liquid sugar for the manufacture of beverages

Beet sugar

The harvested beets are washed and sliced into thin slices called cossettes. The sugar contained in the cossettes is extracted with hot water (diffusion) producing a sugar-rich raw juice. To this raw juice it is added lime (CaO) which reacts with many non-sugar impurities to form insoluble compounds on one hand and to saponify amides and to degrade the invert sugar on the other (Byung and Rex, 1996). Then CO_2 is added which precipitates the excess lime forming $CaCO_3$ carrying with it impurities which then are filtered out. This is the first carbonatation. The filtered juice is further carbonated in order to precipitate any remaining lime and also to precipitate any soluble Ca salts. This is the second carbonatation. The filtered juice after that is called thin juice and contains less than 20% dry substances (DS)[5]. The thin juice is then concentrated by evaporation to produce thick juice which contains about 70% DS. Thick juice is then evaporated in vacuum pans to produce white sugar while the syrup left (green syrup) is put in the process twice more (Figure

---

[5] Dry substances, DS, or refractometric dry solids (RDS) are g solids / 100 g solution. Usually for sugar solutions they are given in degrees Brix (°Bx). Strictly speaking, degrees Brix are grams of sucrose per 100 grams of solution. Sugar in solution affects the specific gravity and the optical properties, especially the refractive index and the rotation of the polarized light. Refractometers have been calibrated to give directly the °Bx. Polarization (Pol) gives the sucrose content of a solution in grams of sucrose per 100 grams of solution measured by the optical rotation of polarized light. The degrees Brix are also used by the wine and fruit juice industries.

3.1). The remaining syrup constitutes the beet molasses. Sugar can be recovered from molasses by adding powdered lime at low temperatures whereby sugar is precipitated as calcium sucrate (Steffen process).

The presence of reducing sugars decreases the solubility of sucrose while the presence of salts tend to increase it. Sucrose solubility in molasses increases with the presence of cations in the following order:

$$K^+ > Na^+ > Ca^{2+} > Mg^{2+}$$
$$1.0 \quad 0.94 \quad 0.66 \quad 0.61$$

The Quentin process (page 114) is based on the fact that K^+ is the most melassigenic cation while Mg^{2+} is the least one and replaces K^+ ions by Mg^{2+} ions using a SAC resin in the Mg^{2+} form.

Sugar inversion during processing results in less sugar. In addition, reducing sugars present some disadvantages. They are more hydrophilic thus retaining more water and give higher moisture to sugar. They also affect caramelization and give a browning effect.

.

The use of IER in beet sugar processing includes:

- Decalcification (softening) of thin juice to avoid the formation of scale in the evaporators
- demineralization of thin juice and/or the exchange of the melassigenic cations K^+ and Na^+ for the less melassigenic Mg^{2+} in the syrup of the second crystallization (Quentin process) in order to increase the sugar yield

- the recovery of sugar and betaine from molasses by chromatographic separation.

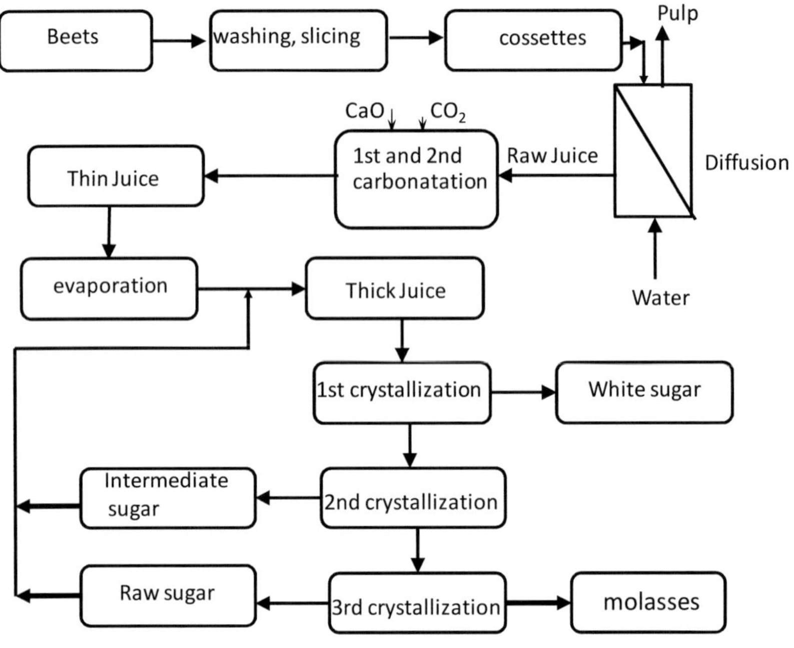

Figure 3.1 Sugar processing in a beet sugar factory

Thin juice decalcification

Thin juice softening, or decalcification since the main hardness ion in thin juice is Ca^{2+}, is achieved using a gel type strong acid cation (SAC) exchange resin in the Na^+ form. During the process, thin juice after the carbonatation step passes through

100

the SAC resin at a flow rate of 20 BV/h and a temperature of about 80°C during which the resin removes Ca^{2+} ions from the solution and replaces them with Na^+ ions fixed initially on the resin, as illustrated in the reaction below:

$$2\ \mathcal{R}\text{-}SO_3Na + Ca^{2+} \leftrightarrows (\mathcal{R}\text{-}SO_3)_2Ca + 2\ Na^+ \qquad (3.1)$$

Some K^+ from the juice is also exchanged with Na^+ of the resin. Regeneration of the resins can be done with a 10% NaCl solution at a level of 200 g $NaCl/L_R$[6] according to the following steps:

- sweetening off with 1 BV of soft water, or until at the exit the effluent has 1 °Bx
- backwash, 0.8 BV soft water
- NaCl injection, 2 BV of 10% NaCl solution at 2 BV/h
- Rinse with 2 BV of soft water at 2 BV/h
- Sweetening on with thin juice until the effluent has 1 °Bx

The operating capacity obtained depends on the hardness of the thin juice. In a thin juice, the monovalent mineral salts K^+ and Na^+ are of the order of 100 meq/L while the hardness ions, Ca^{2+}, are of the order of 5 meq/L. In view of the large excess of monovalent ions compared to Ca^{2+}, the reaction 3.1 above cannot go much to the right. Consequently, the operating capacity of a standard SAC resin in decalcification of juices is relatively low and depends on the hardness level of the thin juice (assuming that the concentration of monovalent cations in the juice remains the same). This is illustrated in table 3.1 below:

---

[6] 200 g $NaCl/L_R$ means 200 g NaCl per liter of resin

Table 3.1 Operating capacity of a SAC resin as a function of the thin juice hardness

Hardness of thin juice (mg CaO/L)	Hardness of thin juice (meq/L)	Resin capacity (g CaO/L_R)	Resin capacity (eq/L_R)
30	1.07	8	0.29
40	1.43	10	0.36
60	2.14	14	0.50
80	2.86	18	0.64
100	3.57	20	0.71
200	7.14	25	0.89

The above capacities hold for a regeneration level of 200 g NaCl/L_R.

The problems associated with this process are that the waste regenerants contain high levels of Cl^- which should be disposed creating environmental problems and that Na^+, a cation that is melassigenic by holding sugar in the molasses thus preventing sugar from crystallizing, are introduced to the thin juice during decalcification. In addition, there is a certain dilution of the thin juice which increases the evaporation cost to a certain extent. Different techniques were developed to address these problems, as described below.

Softening with a weak cation resin

A weak acid cation (WAC) resin in the H form can exchange the H^+ for a cation in the passing solution according to the following reaction:

$$R\text{-}COOH + M^+X^- \rightleftharpoons R\text{-}COOM + H^+X^- \qquad (3.2)$$

This reaction goes to the right only if the acid H^+X^- is weaker than the resin. For example, if the anion X^- is HCO_3^- then the above reaction can go to the right especially if the cation is Ca^{2+}. Thin juice decalcification with WAC resins (Coca *et al*, 2010; Henscheidt *et al*, 1990) is based on the above reaction.

An unwanted side reaction here is sugar inversion at the acidic pH of the resin and of the decalcified juice. During inversion, one mole of sucrose gives one mole of glucose and one mole of fructose.

$$Sucrose + H_2O \quad \xrightarrow{\;H^+\;} \quad glucose \;+\; fructose$$

Since the temperature of the thin juice is around 80°C to avoid bacteria growth, in order to avoid inversion, the flow rate should be high, of the order of 50-100 BV/h and the treated juice, especially at the beginning of the cycle, should be neutralized immediately. WAC resin gave higher operating capacities compared to a SAC resin in the Na^+ form (Coca *et al*, 2010).

The resin is regenerated with dilute sulfuric acid. In order to avoid $CaSO_4$ precipitation the sulfuric acid should have a concentration of about 0.25%. The spent regenerant is recycled back to the diffusion and therefore there is no environmental problem due to waste regenerant.

A variant of this process converts the WAC resin from the H^+ form to the Na^+ form before starting the loading cycle, in order to avoid inversion of sugar (Burkhardt *et al*, 2000; Coca *et al*, 2010).

The NRS process

The NRS process (**N**ew **R**egeneration **S**ystem) is based on the fact that in alkaline conditions sugar forms calcium saccharate[7] which is soluble in water at relatively low temperatures (40°C) (Pannekeet, 1982 and 1984). The regeneration according to this process is done with NaOH dissolved in decalcified thin juice at a temperature of 40°C. The sucrose in the solution with the Ca^{2+} forms calcium saccharate which prevents Ca^{2+} from precipitating in the alkaline solution as $Ca(OH)_2$. The spent regenerant is sent back to the carbonatation step and so there is no waste regenerant neither dilution of the thin juice (fig. 3.2). The IER used in this process is a gel type strong acid cation exchanger. The regeneration level is set at a ratio of the operating capacity to the regeneration level equal to 0.65 (see table 3.2) with a minimum of 32 g $NaOH/L_R$ (0.8 BV of 40 g NaOH/L solution) (Mottard, 1983).

---

[7] Calcium sachharate is the calcium salt of saccharic acid, a dicarboxylic acid. Here, it is in fact sachharated lime also called lime sucrate.

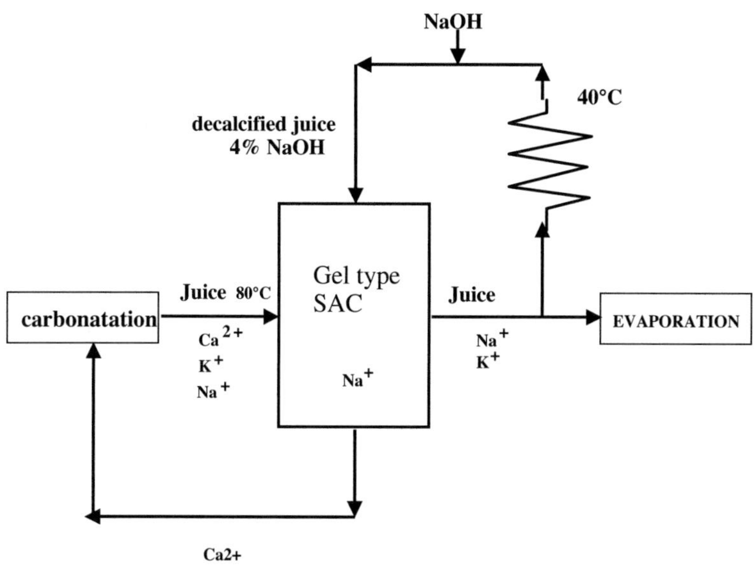

Figure 3.2 NRS Process of thin juice decalcification

Table 3.2. Regeneration level in the NRS process

Hardness in thin juice (mg CaO/L)	Resin capacity (g CaO/LR)	NaOH level for NRS (g NaOH/LR)	Number of BV of 40g NaOH/L solution
30	8	17.6	0.44
40	10	22.0	0.55
60	14	30.8	0.77
80	18	39.6	0.99
100	20	44.0	1.10
200	25	54.9	1.37

For example, assume that hardness is 100 mg CaO/L thin juice. The expected operating capacity from table 3.2 is 20 g CaO/L_R or 0.71 eq/L_R. The 0.65 ratio level of NaOH is then 0.71/ 0.65 = 1.1 eq/L_R or 44 g NaOH/L_R. The regenerant is made by dissolving NaOH in decalcified thin juice at a concentration of 4% NaOH. Therefore, the regenerant level of 1.1 eq/L_R means that 1.1 BV of regenerant will be used to regenerate the resin. The treated volume of thin juice will be 20*1000/100 = 200 BV. At a specific flow rate of 20 BV/h it results into a cycle time of 200/20 = 10 h.

Considering that 100 tons per day of beet result in 120 m3 thin juice per day, a 8000 tons beet per day plant would need to decalcify 400 m3/h thin juice. At a hardness level of 200 mg CaO/L, this means that it will need two parallel columns of resin, 20 m3 each, at 20 BV/h and a cycle length of 6 hours. Alternatively, three vessels in parallel, 2 on loading and 1 on regeneration containing each 20 m3 resin can operate with a cycle length of 12 hours and regeneration time 6 hours. The two columns on loading will be shifted by 6 hours.

The regeneration steps can be described as follows (for the case of upflow, counter-current regeneration):

- drain the thin juice to the level of the resin
- pass upflow decalcified juice at 25 m/h for 2 minutes to fluidize the resin. This step helps the resin to decompact so that the air agitation indicated below is more efficiently done.
- drain the column so that the juice is found about 10 cm above the resin level. Effluent goes to before carbonatation
- air agitate for a few minutes
- pass upflow hot non decalcified thin juice for 15 min at 25 m/h. Effluent goes to filtration after carbonatation.

106

- Cool down the resin to 40°C with decalcified juice upflow at 1 m/h for 20 min.
- Keep decalcified juice flow rate at 1 m/h upflow and inject NaOH so that the NaOH is 4% for at least 1 hour so that the right level of NaOH has passed through.
- Stop NaOH injection and keep the flow rate of decalcified juice for 20 min.
- Rinse upflow with hot non decalcified juice at 1.5 m/h for 2 h.

The regeneration time lasts at least 4 hours

Co-current regeneration is performed similarly except that the cooling down, regenerant injection and rinse are done downflow, co-current.

If non-decalcified juice is used as solvent for the NaOH regeneration instead of decalcified juice, then an operating capacity by about 5% lower will be obtained, compared to decalcified juice (Pannekeet, 1982).

The performance of the resin depends on the regeneration conditions : absence of $CaCO_3$ in the resin, appropriate backwash so that there is no mixing with the silex support (no abrupt introduction of air in the column), use of well filtered juice for backwash etc.

At the end of the sugar campaign, the last regeneration should be followed by the following steps:
- Sweetening off with 2-3 BV of soft water
- HCl wash to clean the resin from any $CaCO_3$ with open valve to let escape the CO_2.
- Conversion of the resin with NaOH to the Na^+ form.
- Store the resin during the inter-campagne in 4% NaOH, or higher concentration if there is a risk of deep freezing.

The Gryllus process

In the Gryllus process, the regenerant is decalcified thick juice (Gryllus and Delavier, 1975; Felber, 1971). This juice contains high concentration of K+ and Na+ and therefore is suitable for regenerating the resin loaded with Ca^{2+}. After the decalcification of the thin juice, it follows the sweetening off step, first with medium Brix juice and then with thick juice. The thick juice regenerant returns to the thick juice followed by sweetening on with thin juice.

In the RDN (ResinDioN) process (Gryllus *et al*, 1998), NaOH is added to the thick juice and this is used as regenerant of the SAC resin. The addition of NaOH in the decalcified thick juice is based on the same principle as the NRS process described above. The spent regenerant returns back to the mother liquor. The Gryllus process can also use the liquor after the second crystallization step as regenerant. After regeneration, this liquor returns to the mother liquor.

In the Gryllus process, it is recommended to use a macroreticular type cation exchanger (converted beforehand to the exhausted K^+ and Na^+ form) because of the high solids content of the regenerant solutions in which the resin works in. At these high solids solutions, gel-type resins shrink too much and the ion exchange reactions take place slowly. In addition, it may cause physical stability problems if gel type resins are used. The macroporosity of the MR resins on the other hand make the access of the functional groups easier. In addition, MR resins are more physically stable and resist osmotic shocks due to alternation of high and low solids solutions. As a rule, for solutions up to about 20-25 Brix one can use gel-type resins, as for example in thin juice, while with higher Brix solutions MR resins are preferred.

Another process of thin juice decalcification is described in a patent (Rousseau 1999 and 2003) where the SAC resin is regenerated with diluted molasses at a temperature around 80°C. In this process the resin is first drained to remove the interstitial amount of thin juice. The process uses one third of 84°Bx molasses diluted down to 75°Bx and then the spent regenerant is mixed with the process molasses to produce a 81°Bx molasses sent to storage.

Thin juice demineralization

Thin juice contains sugar (S) and non-sugar (NS) compounds which include inverted sugar, aminoacids, color compounds and minerals. The sugar content is expressed as g sucrose/100 g solution and non-sugar (non-sucrose) is also expressed as g NS/100 g of solution. It follows that

$$S + NS + W = 100$$

where W is the quantity of water in 100 g of solution. The dry substances, DS, is the sum of S and NS:

$$DS = S + NS$$

The Brix is taken as the grams of DS in 100 grams of juice.

Purity, P, is the percentage of sugar in the dry substances:

$$P = \frac{S}{DS} * 100$$

It was found that the sugar recovery (R) in the crystallization step depends upon the content of the non-sugar (NS) compounds. It is possible to estimate the percent sugar recovery from the purity of the thin juice (P_j) and from the purity of the molasses (P_m) by using the relationship 3.3 (Asadi, 2007). This relationship is based on the fact that the thin juice purity is the same as the thick juice purity (unless there is sugar losses during evaporation) and also that all the NS of the thick juice are found finally in the molasses.

$$R = \frac{(P_j - P_m)}{P_j * (100 - P_m)} * 10000 \qquad (3.3)$$

This relationship is illustrated in figure 3.3 for the case of a molasses purity of 60%.

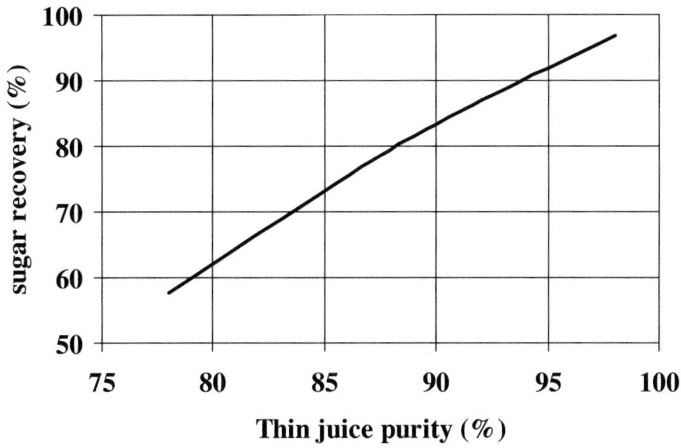

Figure 3.3 Sugar recovery as a function of thin juice purity

In cases where the thin juice purity is low, for example between 80 and 90%, a partial demineralization of the thin juice, for example one third of the thin juice, can result in an increase of the sugar recovery by several percent. Since only part of the thin juice needs to be demineralized, it is preferable to first decalcify all the thin juice and then demineralize part of it before going to evaporation (figure 3.4)

Thin juice demineralization is achieved with a SAC in the H^+ form followed by a WBA in the free base form according to the reactions below.

$$\mathcal{R}\text{-}SO_3H + KCl \rightleftharpoons \mathcal{R}\text{-}SO_3K + HCl \qquad (3.4)$$

$$\mathcal{R}\text{-}N(CH_3)_2 + HCl \rightarrow \mathcal{R}\text{-}N^+(CH_3)_2H\ Cl^- \qquad (3.5)$$

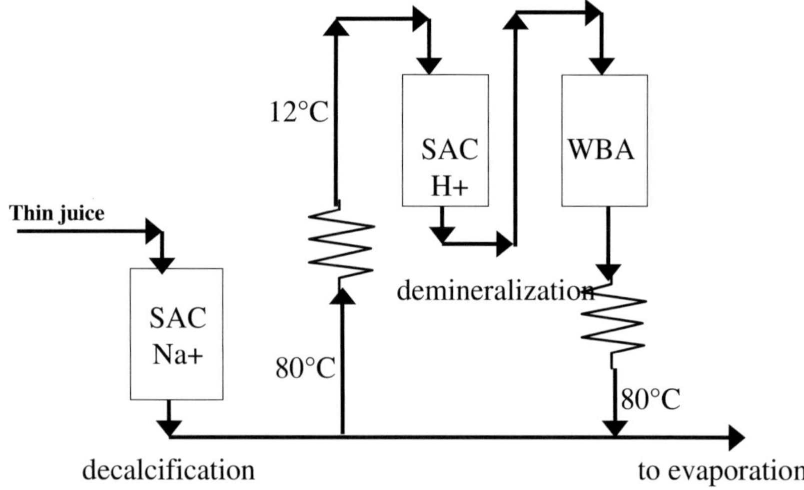

Figure 3.4 Thin juice demineralization

Thin juice passes through the SAC resin at a flow rate of about 2-3 BV/h at 12°C. Because the pH inside the SAC resin as well as of the decationized juice is very acidic, the juice needs to be cooled down to avoid sugar inversion.

Since IER are characterized with a capacity expressed in equivalents per liter resin (eq/L_R), it is practical to express the NS impurities in eq/L thin juice. A possible analysis of a thin juice gives that 100 g of NS contains: 160 meq Na^+, 290 meq K^+, 25 meq Ca^{2+}, 15 meq others and 215 meq betaine, a total of up to 750 meq.

For a thin juice of 15 % DS having a density of 1.055 and a purity of 85%, the above composition can be expressed as:

Na$^+$	35 meq/L
K$^+$	65 meq/L
Ca^{2+}	5 meq/L
Betaine	45 meq/L
TOTAL	150 meq/L

Consider 1 m3 of a thin juice having 15 % DS (or 15 °Bx) and 85% purity. Using 1.055 as the density of this juice, the 1 m^3 is 1.055 tons of thin juice containing 0.158 tons DS, 0.134 tons sugar and 0.024 tons of NS. The operating capacity of the resin to a 5% leakage end-point is about 200 g NS/L$_R$ (approximately 1.5 eq/L$_R$). This means that the resin can remove 95% of the NS of about 8.5 BV of thin juice. The purity of the demineralized juice will be calculated from the sugar content and the 5% NS leakage. The result is about 99%. If for example the desired purity of the thick juice before crystallization is 95%, then 70% of the thin juice can be demineralized and mixed with the thin juice of 85% purity before evaporation.

Taking a specific flow rate of 3 BV/h the loading cycle time will then be 8.5/3= 2.8 h. If for example we have 100 m^3/h of thin juice, this juice will first be decalcified with 5 m^3 of a SAC resin (20 BV/h) then a 70 m^3/h stream will be demineralized using 23 m^3 of SAC resin and 23 m^3 of a WBA resin then mixed with remaining 30 m^3/h juice to send to evaporation.

Regeneration is done with 120 g H$_2$SO$_4$/L$_R$ for the cation and 80 g NaOH/L$_R$ or 70 g NH$_4$OH / L$_R$ for the anion resin. If NH$_4$OH is used, the residues of the spent regenerants can be used in fertilizers production.

After the SAC, the pH becomes acidic and in order to avoid sucrose inversion the juice should be cooled down to 10-12°C. Gel-type SAC resins and styrenic or acrylic type WBA resins can be used. Acrylic WBA being a little stronger base than styrenic, they raise the pH somewhat higher which can be an advantage of these resins.

The "new demineralisation process" is a process developed jointly by Générale Sucrière and Rohm and Haas Company where thick juice was demineralized by a mixed bed of a WAC resin and an MR type acrylic WBA resin. At the same time, considerable decolourization was also obtained. The process did not commercialize at that time for economical reasons, essentially because the life time of the WBA resin was found to be too short (Rousseau *et al*, 1990).

The Quentin process

The principle of the Quentin process is to treat the second strike syrup and replace the highly melassigenic cations K^+ and Na^+ by the less melassigenic Mg^{2+} ions so that less sugar goes with the molasses.

$$(\mathcal{R}\text{-}SO_3)_2Mg + 2\,K^+ \rightleftarrows 2\,\mathcal{R}\text{-}SO_3K + Mg^{2+}$$

The syrup is treated at 90°C and at about 2 BV/h. A macroreticular SAC resin is used converted to the Mg^{2+} form using a $MgCl_2$ solution. This process is used in the central Europe area where Mg salts are available as by-products of the potash industry.

Betaine

Betaine is an aminoacid existing in zwitterionic form at neutral pH with a positively charged quaternary ammonium group and a negatively charged carboxylate group. The pKa of the carboxylic group is 2.3.

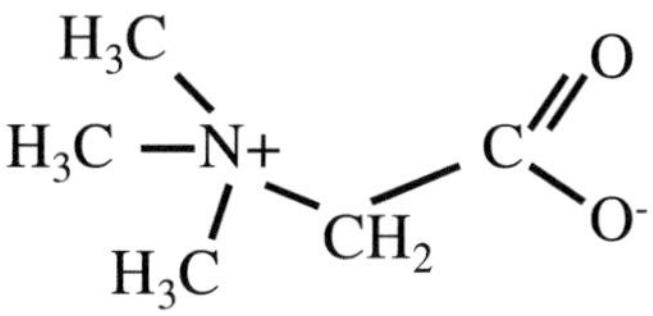

Figure 3.5 Chemical structure of betaine

Beraine was discovered in beet sugar. It has become popular recently in supplements for its potential benefits in heart decease, promoting muscle gain and fat loss. During standard sugar processing, betaine is found in the molasses.

Among the earliest industrial applications of ion exchange was in betaine recovery (Bennett, 1945). In this process, betaine was recovered from the waste waters of the Steffen process, after the removal of sugar from molasses with lime at low temperatures. The waste waters passed first through a SAC resin in the H^+ form which removed all cations plus betaine. Since the resin had higher affinity for inorganic cations, K^+, Na^+, than for betaine, the betaine leaked first and the effluent was fractionated to a betaine-rich fraction and a betaine-poor fraction. Betaine was recovered either from the acidic effluent of the SAC resin, or the effluent of the SAC resin was passing through a WBA resin which removed the acids and betaine was recovered as neutral molecule by crystallization or extraction with an alcohol.

Today betaine is recovered either from molasses, from the rest molasses after desugarization using the Steffen process, or from vinasse. Molasses desugarization using SAC exchange resins in K^+ form, or mixed cation form, is described later, page 140. It is a chromatographic separation of sugar from non-sugars based on ion exclusion. The eluent is water. In the eluate, they are recovered first the non-sugars (raffinate), followed by the sugar and last a betaine-rich fraction (Heikkila *et al*, 1985). In fact, using a SMB there are separated the non-sugars from the sugar+betaine which then is further separated by a coupled SMB to sugar and betaine (see also page 94).A possible flowsheet for betaine recovery is illustrated in figure 3.6.

Weak acid cation exchange resins in the H^+ form have been proposed for chromatographic separation of betaine from sugar but also from additional compounds to betaine and sugar such as erythritol, inositol, raffinose, mannitol and carboxylic acids (Paananen *et al*, 2007). Temperature is 65-90°C and the feed solution pH is adjusted to 3-4.5 in order to keep the WAC resin in the H form. The feed solution pH affects in fact the retention time of betaine and the other carboxylic acids in the feed solution. The lower the pKa value of the carboxylic acids in the feed, the shorter is the retention time, except for betaine.

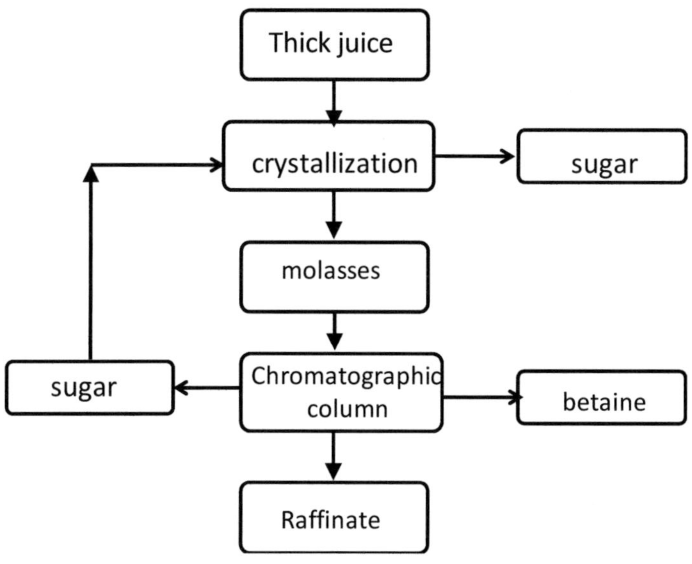

Figure 3.6 Possible flowsheet for betaine recovery

Cane sugar

The processing steps of sugar production from sugarcane is illustrated in figure 3.7 below:

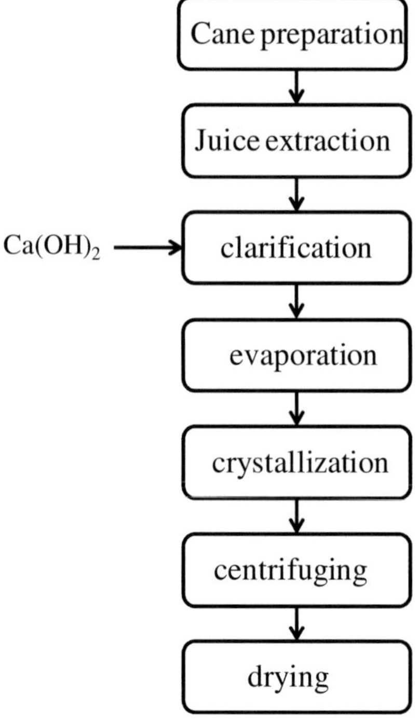

Figure 3.7

After sugar has been extracted from the cane by diffusion with water or by milling, the juice is clarified by heating it with lime. The pH of the raw juice is acidic, 4-4.5, and the addition of lime to a pH of about 7 among others it prevents the inversion of sugar (Laksameethanasan *et al*, 2012). The main reaction in this simple defecation process is between lime and the phosphates content of the juice causing the precipitation of calcium phosphates. Impurities are entrained out with the insoluble calcium salts, settle down and the supernatant sugar solution goes to evaporators. After evaporation, the juice has about 65%

118

solids. It is filtered and sent to the vacuum pans for crystallization. Sugar is recovered in three crystallization stages, called A, B and C. A centrifuge separates the sugar crystals from the mother liquor. The sugar from A centrifuge is washed, dried and sold as raw cane sugar. The sugar from the B and C centrifuges is remelt and sent to the evaporation station. The remaining liquor from the third crystallization is the molasses. Because simple defecation does not remove all the impurities, raw sugar has a polarization of 94-97%, ash > 0.4% and a color that can be from 500 to >3000 ICU. Some cane sugar mills use carbonatation or sulfitation to produce a so called plantation white sugar by adding CO_2 or SO_2 to the lime. The addition of SO_2 prevents further color development in the process.

Raw sugar is different from brown sugar which is refined white sugar mixed with molasses to get the brown color. Some sugar mills have back-end refineries where part of the raw sugar produced is refined to produce higher quality white sugar.

Although the main steps in recovering cane sugar are similar to those of beet sugar, there are some differences that may have an effect on the ion exchange resins plant.

The composition of the clarified juice of the cane sugar differs from that of beet sugar in that it contains fibers and also it contains more invert sugar and more phenolic compounds that give a dark colour to the juice. During the purification step the fibrous materials are removed by filtration and then lime is added while temperature is raised which neutralizes the natural acidity, thus stopping the further decay of sucrose to glucose and fructose, and also to facilitate the precipitation of impurities such as proteins. The clarification step is not as complete as the purification of the beet sugar raw juice and leaves more invert sugar which is not decomposed as is the case in beet sugar juice.

By filtration, a clear juice is obtained which then goes to evaporation and crystallization in three stages. The raw sugar is then sent to local consumption or exported to refineries.

Contrary to the beet sugar processing, there is no decalcification with IER in cane sugar mills because the life time of the resins is very short due to the impurities of the clear juice. In recent years Applexion® (Novasep) has developed a process where clarified juice undergoes microfiltration after which the juice is decalcified before crystallization thus producing sugar of a better quality than conventional: fewer colours, less ash and faster crystallization speed.
Processes to produce white sugar directly in a sugar mill rather than at refineries have been developed, motivated by the cost savings involved. These processes are described later in this chapter (page 129).

Cane sugar refining: decolorization

During processing of cane sugar, a number of color compounds are formed by heat or enzymes such as caramels and various products of the Maillard reaction including melanoidins. The Maillard reaction is the reaction of the amine groups of aminoacids with the carbonyl group of reducing sugars[8] forming a mixture of poorly characterized products, responsible for taste,

---

[8] Reducing sugars are those sugars that have an open chain structure with an aldehyde group. This aldehyde group can be oxidized via a redox reaction while another compound is reduced. Ketone groups cannot be oxidized without decomposition of the sugar but sugars with ketone groups in their open-chain structure can form aldehyde groups by isomerization. Therefore, sugars like fructose are considered as reducing sugars.

120

odor and color formation. Generally speaking, the colorants can be classified to ionic-aromatic, ionic-nonaromatic, nonionic-aromatic and nonionic-nonaromatic (Chen, 1985). The ionic compounds bear weak anion groups (carboxylic groups). The aromatic nonionic include polyphenolic compounds. The nonaromatic contain conjugated double bonds. In the refining of raw cane sugar, all these compounds are removed to a certain extent to produce white sugar.

The raw sugar from the mill is first washed with a saturated syrup to remove a thin film of molasses that is found on the crystals (affination). The mixture of crystals and syrup, called magma, contains about 10% water and has a temperature of 45-70°C. To avoid sugar inversion, lime may be added to bring pH above 7.2. Then the crystals are separated from the syrup by centrifugatin and washed with hot water. The wash water is mixed with the syrup and is called affination syrup which is then used to the affination step. After washing and further centrifugation, the crystals are sent to a melter where they are dissolved in water. The resulting affination liquor has a slightly alkaline pH, between 7.2 and 8.0, a solids concentration between 62 and 68° Brix and a temperature 80-85°C. The raw washed sugar liquor contains insoluble materials such as clay, sand, finely dispersed particles and colloids. It is clarified by phosphatation where lime and phosphoric acid are added to form calcium phosphate precipitate or carbonatation where $CaCO_3$ precipitate is formed by the addition of lime and CO_2. Much of the suspended matters, as some of the colorants, are trapped by the flocculent. Depending on the sugars and the operating conditions, a 25-40% color removal can be achieved with phosphatation and 30-50% by carbonatation (Chen, 1985). The flocculent is filtered and sent for decolorization. The decolorization results in a fine liquor from which the sugar is crystallized, centrifuged and dried to produce white sugar.

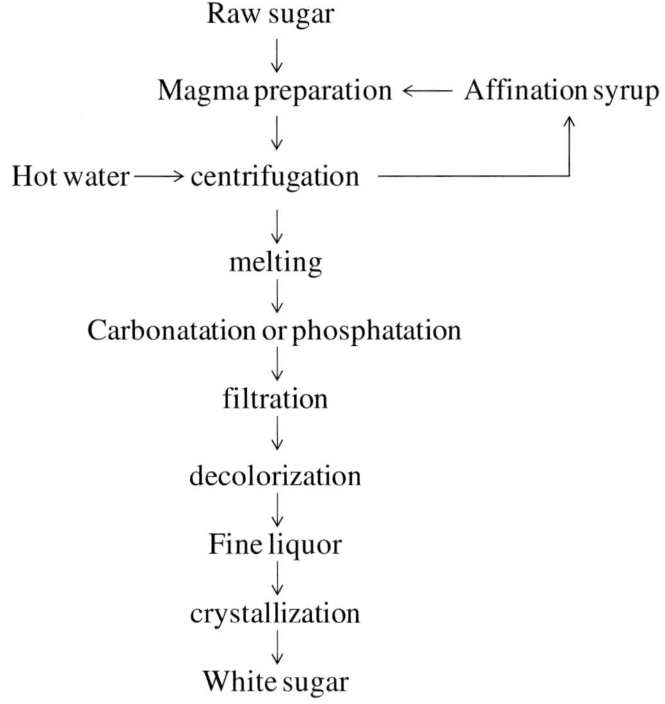

Figure 3.8 Cane sugar refining

The Talofloc process is a simultaneous clarification and decolorization process. A cationic surfactant is used to react with the weak anionic groups of the colorants and thus precipitating it followed by phosphatation where the precipitated colorants are removed with the calcium phosphate flocculent. The surfactant is of dialkyl dimethyl quaternary ammonium compounds. The dioctadecyl dimethyl ammonium chloride is available under the name of Talofloc.

The techniques used to decolorize sugar are:

Carbon adsorbents (bone char, granular carbons and powdered carbons)

Ion exchange resins and polymeric adsorbents

Bone char is regenerated and reused many times. Granular activated carbon is reused about 25 times. Powdered carbons are used only once or twice.

A study was performed some years ago at the Marseille (France) cane sugar refinery (Celle and Hervé, 1980) where the economics of the decolorization process using bone char, activated carbon and ion exchange were compared. As a result, this refinery switched from bone char to ion exchange and operates successfully since then. At about this time, the Huletts Sugar refinery in Durban, S.Africa switched from bone char to ion exchange. In fact, ion exchange technology is increasing in use because of the easier regeneration (no heat necessary and therefore savings in energy costs).

The ion exchange resin used is an MR type SBA resin in the Cl^- form, with either acrylic or styrenic matrix. Most frequently, type 1 styrenic SBA exchangers are used even though type 2 can also be used. Gel type SBA resins, especially acrylic, have also been used in some geographical areas as a lower cost, less performing and with shorter life time material. The mechanism of color removal by SBA exchange resins is adsorption and ion exchange at the same time. The ion exchange mechanism is analogous to the Talofloc process described above.

Generally speaking, the acrylic resins are more hydrophilic than styrenic. MR type acrylic SBA resins remove somewhat less color and they regenerate more completely than MR styrenic resins thus having longer life time.

Depending on the colour of the feed solution and the desired colour of the treated juice, only one column of resin can be used for decolourization or two columns in series can be used, the second one acting as a polisher. Contrary to the traditional merry-go-round system where at the end of the loading cycle the polishing column becomes head column and the freshly regenerated column becomes polisher (page 79), here the first column remains first column and the polishing column remains polishing column for the whole life time of the resin of the head column. The two columns in series work therefore as a pair. When the resin in the head column reaches the end of its life, the polishing resin goes to the head position and new resin is installed in the polishing column. In this way, the resin in the polishing position does not treat high-colour juice and is maintained in a relatively clean condition giving a better decolorization to the treated juice.In some plants where two columns in series are used, the first column is loaded with acrylic resin and the second with styrenic so that the easier regenerated acrylic resin in the head column protects the styrenic resin from fouling. Regeneration is done from the second column towards the first one.

The decolorization achieved with ion exchange resins depends on the feed composition and the operating conditions such as cycle length, regenerant level and composition or backwash conditions. As an indication, with used styrenic resins in the middle of their life time, with one pass approximately 65% decolourization is obtained, with two passes 85% decolourization is obtained and with three passes 90%. When the resins are new, two passes decolorize better than 90%.
The color units used are the International Color Units (ICU) and are calculated using the following expression:

124

$$ICU = 1000*A/b/c$$

Where A is the absorbancy at 420 nm, b is the absorbing path in cm and c is the solution concentration in g/cm3.

Two columns in series of acrylic resins, with new resin in the polishing column, decolorize about 70%, 65% at half life time and 50% at end of life of the lead column (about 350 cycles, half as a polisher and half as lead column). Typically, the flow rate is 2-4 BV/h (1 BV is the resin volume of one column), the feed juice has 60-65°Bx and the temperature is 70°C for 60°Bx and 80°C for 65°Bx juice. The temperature and the Brix will determine the viscosity of the syrup and this will determine the pressure drop at a certain linear velocity. The color of the feed solution is usually about 700-1000 ICU.

The decolorization obtained with the anion exchanger depends among others on the mineral salts (ash) concentration in the syrup. Color materials in the syrup are fixed on the resin by electrostatic interaction (ion exchange) and by hydrophobic interactions. Salts present in the syrup compete for the ion exchange sites with the color bodies and decrease the degree of decolorization. In fact, high concentration salt solutions are used to regenerate the resin and elute the absorbed color[9].At the same time, because of this ion exchange between the Cl⁻ ions fixed on the resin and the anions in the syrup, including the color material, the Cl⁻ concentration in the decolorized syrup increases along with the conductivity. In some cases, the pH of the

---

[9]This should not be confounded with the "salting-out effect" where the adsorption of a non-dissociated or little dissociated molecule increases when the external solution contains high salt concentration, due to the decreased solubility of this molecule in the salt solution.

decolorized syrup drops down to 4-4.5 from an average value of 6 of the feed syrup. This pH decrease is not experienced always and depends on the feed syrup composition. New resins present this characteristic more than used resins.

The quantity of color loaded on the resin depends on the color content of the syrup to decolorize. The equilibrium between color on the resin and color in solution is however very long to reach. In practice, the cycle stops well before resin saturation. The cycle length depends on the feed colour and is stopped when the resin is loaded with a certain quantity of color matter. This is in order to avoid a premature fouling of the resin because the more color the resin is loaded with, the more it remains on the resin after regeneration and the resin becomes fouled. An approximate indication of the colour to load on the resin during the loading cycle based on practical experience is that *the number of kg of DS passed through one liter of resin multiplied by the color of the syrup in ICU should be about 40000 kg $DS*ICU/L_R$.* This applies for syrups with color of 500-1000 ICU. For example, if the feed solution has 60°Bx and a density of 1.3 then the dry solids are DS = 60*1.3/100 = 0.78 kg/L. If the resin treats β liters syrup (that is, β BV), then there will be treated 0.78*β kg/L_R. Then, if the feed color is 1000 ICU, then the color passed from the resin is (DS*ICU*β) kg*ICU/L_R and this number should be less than 40000. Consequently, the loading cycle should be limited to β = 40000*100/(1000*1.3*60)=51 L/L_R (=BV) where the BV refers to the resin volume in one column. The expected life time of the resins is about 300-350 cycles. For syrups with color higher than 1000 ICU it is better to use acrylic resins which as mentioned above regenerate easier or combination of acrylic (in the head column) and styrenic (in the polishing column).

126

A mixture of 10% NaCl and 0.2-0.5 % NaOH is used as regenerant. If two columns in series are used, the regenerant passes first from the second column then to the first. The regenerant level is 200-250 g $NaCl/L_R$ where the L_R refers to one column. If only one column is used, the regenerant level is about 150 g $NaCl/L_R$. This regenerant composition assures that the form of the regenerated resin is the Cl^- form. It has been found that a higher NaOH content allows a better regeneration efficiency, however if the NaOH content exceeds the 0.2-0.5% the life time of the resin decreases significantly because the resin is converted to a certain extent to the OH form and due to the high temperature the strong base functional groups undergo the Hofmann degradation and are converted to weak base groups.Weak base resins decolorize significantly less compared to strong base resins. The 0.2-0.5% NaOH in the 10% NaCl solution ensures that the fraction of the strong base groups found in the OH^- form after regeneration is kept to very low level and at the same time it elutes more efficiently the color bodies absorbed on the resin.In any way, the OH^- groups are quickly exchanged with the anionic groups in the syrup (ash, color) so that the functional groups of the resins are not found in the OH^- form for long during the cycle.

Every about 15 cycles, it is advantageous to clean up the resins using 2 BV of a 1.5-2% HCl solution.
The process steps in cane sugar decolorization are as follows:
- Service cycle, 2-4 BV/h, 60-68°Bx, temperature about 80°C.
- Sweetening off with 2-3 BV of hot water to displace the syrup.
- Backwash to remove foreign material and to classify the resin.

- Drain the water to about 20-30 cm above the settled resin bed.
- Regenerant injection at about 2 BV/h or 1 hour contact time at 50-60°C.
- Slow rinse at the same flow rate as the regeneration with 2 BV of water
- Fast rinse at 10 BV/h for 30 to 60 minutes.
- Sweetening on

The disadvantage of the decolorization process with IER is the disposal of the waste regenerants. Recent techniques however have been developed that minimize wastes and recycle the regenerant.

In one of them (Bento, 1999) the IER are regenerated with a sucrose solution containing calcium hydroxide and a small quantity of a chloride salt like $CaCl_2$ or $NaCl$ at a temperature of 40-70°C. 2 BV of this regenerant is used at a flow rate of 2-4 BV/h. The sucrose in solution increases the solubility of the $Ca(OH)_2$. After regeneration the Ca^{2+} ions can be removed as $CaCO_3$ by the addition of CO_2 or Na_2CO_3 and filtered off. The spent regenerant can then be recycled back to process.

In another process (by Novasep Applexion® or by GEA Process Engineering and Escon Engineering Services and Consulting GmbH) the spent regenerants from the NaCl+NaOH regeneration are treated with nanofiltration where the retentate contains more than 90% of the color and the permeate is then used to make the rest of the spent regenerant after the addition of the appropriate quantities of NaCl and NaOH. In this way, the brine regenerant is recycled and the waste regenerants are reduced considerably.

Due to economical advantages, processes for direct production of white sugar at the sugar mills without producing raw sugar which is then sent to refineries, have been developed.

One process for white sugar production, the White Sugar Mill (WSM) Process, was developed in South Africa, jointly by Tongaat-Hulett Sugar Ltd and South African Bioproducts Ltd (Fechter *et al*, 2001; Rossiter *et al*, 2002). It involves ultrafiltration, demineralization and decolorization of cane sugar juice. Figure 3.5 gives the flowsheet of the WSM production at the Felixton mill of Tongaat-Hulett Sugar.

In this process the clear juice is evaporated to 25°Bx, filtered and then treated with UF membranes at 90°C to remove high molecular weight substances that risk fouling the resins. The permeate is then cooled down to 10°C, to minimize sugar inversion, and treated by ion exchange. The IER treatment consists of demineralization followed by decolorization. Demineralization uses two carousel units of rotating resin columns, equipped with a multi-port distribution valve developed by Tongaat-Huletts Sugar. One of the units is for the SAC resin and the second for the WBA resin. The ultrafiltered juice passes through the SAC and the WBA resin and then back and forth through more SAC and WBA resin columns in a predetermined number of passes, generally three. Regeneration of the deminaralization units was performed with HCl or

HNO_3[10] for the SAC resin and NH_4OH for the WBA resin. The waste effluents are concentrated and NH_4NO_3 recovered and used as fertilizers.

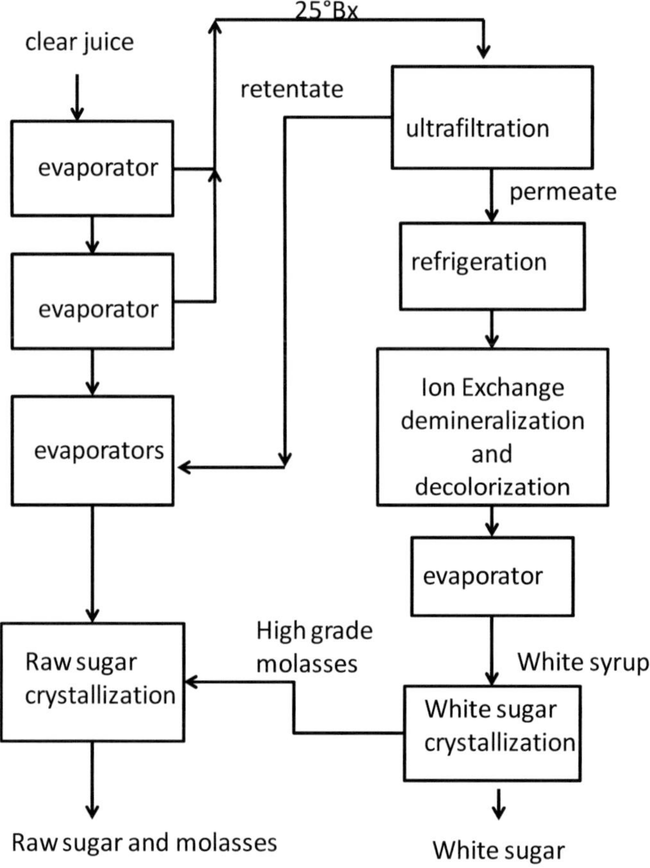

Figure 3.9 Flowsheet of WSM production at the Felixton mill of Tongaat-Hulett Sugar (Courtesy of Tongaat-Hulett Sugar Ltd).

---

[10]safety measures are recommended by the resin manufacturers when HNO_3 is used on ion exchange resins

130

Following demineralization, the juice is decolorized on another carousel using a SBA resin in the Cl⁻ form.Acrylic type SBA exchange resin was chosen. The decolorization unit reduces the color from 2500 ICU of the deashed juice down to 400 ICU.The decolorization resin is regenerated with 10% NaCl solution containing 0.2-0.5% NaOH, as described in the previous paragraph.

In another process developed at the Audubon Sugar Institute (Rein *et al*, 2007),white sugar is produced at the mill without the use of ultrafiltration. Clarified juice is treated with granular activated carbon (GAC) columns for decolorization followed by a SAC in the H^+ form and a WBA resin in the free base form for demineralization. In addition, oxidative agents such as H_2O_2 have been added to increase decolorization and to oxidize coloring molecules, helping to remove them by the ion exchange resins. Before the SAC resin, the juice is cooled down to 10°C to prevent sugar inversion due to the low pH in the SAC resin. The juice is then reheated and sent to evaporators where the concentration of sugar reaches the 60%. This juice is then purified to reduce turbidity and then it goes to crystallization.

In the DRD (Dedini Refinato Direto, Dedini Direct Refined) process (Olivério and Boscariol, 2006) clarified juice goes to evaporation where the juice attains the 60-65°Bx. After the last evaporation, the juice is clarified and filtered before going to the ion exchange system. The IER in the DRD process consists of three columns in series, the first being a decalcification unit while the last two are decolorizing columns. Regeneration is performed with alkaline NaCl solution.

With this process, the input syrup colour was on the average 8776 ICU while on the average the output syrup colour was

4216 ICU or an average of 52% decolourization. The crystallized sugar had about 40 ICU color. The resins which were not disclosed showed relatively low degradation of performance after 48 cycles in spite of the high color load per cycle.

Liquid sugar

Liquid sugar is a term covering a range of products from liquid sucrose solutions of about 67°Bx to total liquid invert sugar at 77 °Bx. Liquid sugars are widely used in the beverage industry and other food products in most countries except in the United States where high fructose syrups are mainly employed. The reason for this is essentially financial, the cost of liquid sucrose is high compared to the High Fructose Corn Syrup (HFCS) in the US due to Government policies while in the rest of the world this is not the case.

Liquid sucrose comes in general in 67°Bx solutions and it uses beet or cane sugar as raw material. Depending on the quality of the crystalline sugar used as raw material, the liquid sugar solution made by dissolving the crystalline sugar has a color in the range of 50 to 500 ICU or more and an ash content of 0.05 to 0.4 % of the DS or more. The specifications for the liquid sugar for colour and ash are defined by the end users. For high quality liquid sugar it is around 35 ICU color and 0.01% ash content. Therefore, the treatments with ion exchange vary according to the raw material and the customer specifications. Some flow sheets with IER are the following:

1) crystallized sugar → remelt → decolourization →GAC

2) crystallized sugar → remelt → decolourization → mixed bed demineralization→GAC

3) crystallized sugar → remelt → decolourization → deanionization → decationization → GAC

If the turbidity is high, there may be a membrane filtration before the IX. In addition to the above IER columns, the flow sheets for the liquid sugar manufacturing can include an activated carbon polishing.

Decolorization is obtained using MR type SBA exchange resins in the Cl⁻ form, similar as in cane sugar refining discussed before. Styrenic resins give a higher decolourization than acrylic resins here. Single pass, that is one column, can be used, or double pass, two columns in series, or triple pass, depending on the degree of decolourization we want to achieve. In order to obtain low color syrup, the load on the resin should be kept at lower levels than in the decolorization of raw cane sugar discussed in page 105 and where the colour in the feed solution was in the range of 500-1000 ICU. With a feed at 50-200 ICU, with single pass approximately 70-80% decolorization is obtained while with double pass 85-90%. For example, with 200 ICU in the feed solution, double pass would bring color down to 30 ICU. With 100 ICU in the feed, single pass would give the same result.

In a similar way as for decolorization at the refinery, in order to maintain a low level of color leakage, the colour load during the service cycle expressed as the number of kg dry solids per liter resin multiplied by the color in ICU should be kept at about 15000. For example, for a syrup with 120 ICU and 67°Bx (density 1.3), the volume to treat would be 15000*100/(120*1.3*67) = 144 BV. Typical operating conditions are: Flow rate 2-4 BV/h (1 BV refers to the resin volume in one column) and temperature 80°C.In the above example, at 4 BV/h the cycle length is therefore 144/4 = 36 hours.

If low ash level of the liquid sucrose is required, the solution can be demineralized after decolorization. With low ash and color in the feed solution, this can be done with a mixed bed of a WAC in the H form and a SBA resin in the OH⁻ form, with a cation to anion resin ratio of 1 to 2. The solution passes through the resin bed at a flow rate of 2 BV/h and at a temperature of 50°C. Operating capacity is about 0.3 eq/L_R where Liter of Resin refers here to one liter of the mixed bed. Assuming that the equivalent weight of "ash" is that of NaCl, that is 58.5, then the operating capacity of the mixed bed is 0.3*58.5=17.55 g ash/L_R. With a feed solution with ash content of 0.2% of DS and 60°Bx and a density of 1.3 one obtains an ash content of 0.2*60*1.3/1000=1.56 g ash/L solution. Therefore, the cycle length will be 17.55/1.56 = 11.25 BV or 11.25/2 = 5.625 hours. Regeneration is done with 70 g HCl/L_R for the WAC and 80 g NaOH/L_R for the SBA resin. The mixed bed can reduce the ash content of the feed solution down to less than 0.01 % of DS. Care must be taken in mixing the cation and anion exchangers components of the mixed bed after regeneration so that the WAC being heavier than the SBA resin does not settle first and forms a layer at the bottom of the column. The resin beads of the two components of the mixed bed are in close contact and this minimizes the time that the sugar is found in the acidic environment of the WAC resin. This ensures a very low level of inversion.

Another way to demineralize the sugar solution and to avoid inversion has been reported (Lanxess; Purolite 2009) with reverse demineralization, that is, having first a SBA resin in OH⁻ form followed by a WAC resin in H form. The SBA resin removes the anions while the WAC fixes the cations which now are found in alkaline pH and neutralizes the solution. Of course this arrangement is feasible for the third flow sheet shown above, where there should be no significant concentrations of alkali earth elements that risk precipitating in alkaline pH. This arrangement prevents the inversion of sugar at the low pH of the WAC resin. Because of the higher capacity of the WAC resin, the volume of the SBA resin in the first column is double of that of the WAC resin in the second column. Loading flow rate is 2 BV/h for the SBA resin and therefore 4 BV/h for the WAC resin and temperature is 50°C. Regeneration levels are 80 g NaOH/L_R for the SBA and 70 g HCl/L_R for the WAC resin.

If the raw material has very high color and ashes, like with raw sugar, decolorization and/or demineralization should be done in two or even three stages in order to meet the liquid sugar specs.

There is an interest in using liquid invert sugar in the beverage industry for two reasons. First, it comes at 76°Bx instead of 67°Bx of liquid sucrose so there is less water to ship and second it has longer microbiological stability. Sucrose hydrolyzes at low pH to one molecule of glucose and one molecule of fructose and this reaction is called inversion. It can be achieved by the addition of an acid like HCl, by the invertase enzyme or by ion exchange. SAC resin in the H^+ form can act as a solid acidic catalyst to invert sucrose to glucose and fructose. A typical product for that purpose is a low crosslinked gel-type SAC resin in H^+ form. The sucrose solution passes through a column with this resin at a certain temperature and flow rate. The degree of inversion depends on the Brix of the solution, on the temperature and on the flow rate of the syrup through the resin bed. If the syrup is not pure but contains salts, the acidic catalyst will fix the cations and will progressively lose its catalytic properties. Therefore, the syrup has to be decationized first. In production of liquid invert sugar, the flow sheet employed is a SAC to decationize the juice followed by a SAC for inversion, followed by a WBA for removing the anions.

A side reaction that takes place is the formation of hydroxymethylfurfural (HMF) from the dehydration of fructose at low pH (Shaw *et al*, 1967).

HMF

HMF is a colour precursor and therefore this side reaction should be avoided. This is achieved by using the appropriate temperature. It is suggested to use a temperature of 40°C or less to avoid formation of HMF.

Figure 3.10 illustrates the relationship between operating conditions and extent of inversion. If for example for a 60 Brix syrup a temperature of 40°C is chosen for inversion then in order to have an inversion of 95% a specific flow rate of 1 BV/h

is needed while for 50% inversion the required specific flow rate is 4 BV/h. Syrup of 60°Bx has a viscosity at 40°C of about 20

136

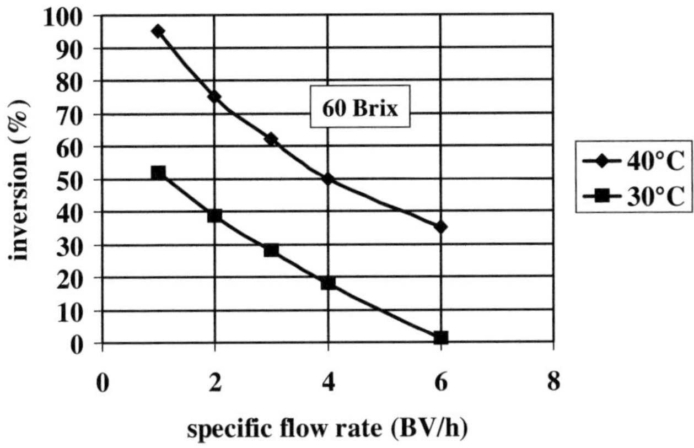

Figure 3.10 Effect of specific flow rate on inversion

cp (centipoise). At this viscosity a typical low crosslinking gel SAC shows a pressure drop across the resin bed as indicated in figure 3.11.

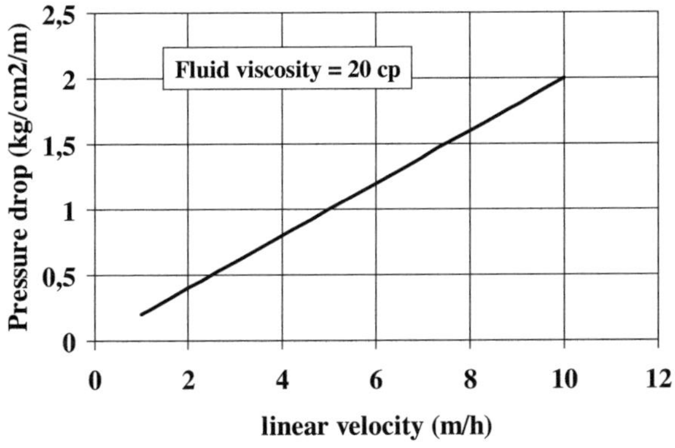

Figure 3.11 Pressure drop vs linear velocity for an inversion resin in a 60°Bx syrup at 40°C

Figure 3.11 should be used to decide the column dimensions. For the case where 95% inversion was chosen, where a 1 BV/h should be applied, if the column is designed so that the bed height of the resin is 1.5 meters, then the linear velocity of the syrup will be 1.5 m/h and the pressure across the resin bed will be 0.45 kg/cm². For the case where 50% inversion is searched, 4 BV/h means 6 m/h and the pressure across the resin bed becomes 1.8 kg/cm². If this pressure is too high, a larger diameter column should be taken resulting into a lower bed height and lower pressure.

For a liquid invert sugar with a raw sugar of relatively good quality (50-200 ICU, 0.2% ash), the following flow sheet can be used:

$$SAC(H^+) \quad \rightarrow \quad SAC\,(H^+) \quad \rightarrow \quad WBA$$
$$\text{Decationization} \quad \text{inversion} \quad \text{deacidification}$$

The first SAC resin removes all the cations from the solution and replaces them with H^+. At the same time inversion starts at this resin. An MR type resin should be used here due to the high DS of the solution. The inversion resin is a low crosslinking density gel type SAC resin and the deacidification resin is an MR styrenic WBA resin. Since the inversion resin acts as a catalyst for the inversion and it does not undergo frequently sequences like loading, rinse, regeneration etc, it does not experience osmotic shocks that risk breaking the resin.

Take for example a 10 m^3/h sugar syrup at 60 Brix and ash of 0.1% of DS of which we want to invert to 95%. In order to avoid HMF formation, temperature should be kept at 40°C and because of the high viscosity of the syrup at this temperature

138

(about 20 cp), linear flow rate should be kept at or below a level that gives a pressure on the resin bed of 0.5 kg/cm² (about 0.5 bars). From figure 3.5 we see that the specific flow rate on the inversion resin should be 1 BV/h and this gives 10 m^3 resin volume. In order to have the above specified pressure on the resin bed or less, the column diameter should be 3 meters which would give a resin bed height of 1.4 meters and a linear velocity of the syrup of 1.4 m/h. From figure3.11 we have that the pressure drop would be about 0.28 kg/cm²/m or 0.4 kg/cm² total pressure on the resin bed.

The decationization resin can accept somewhat higher linear velocities because the corresponding pressure drop curve would be lower than that of the inversion resin, first because the resin is "harder", it is an MR resin with higher DVB level, and second because some resin grades come with a larger particle size. For example, we can have a specific flow rate of 2 BV/h and a 2.2 meters diameter column giving 5 m^3 of resin, 2.6 m/h velocity and a bed height of 1.3 m. Pressure drop is expected to be about 0.38 kg/cm²/m or a total pressure of 0.5 kg/cm². Similar considerations apply for the WBA resin.

With an operating capacity for the two demineralization columns taken as 0.6 eq/L_R and with an ash content of 0.1%, or 13.3 eq/m^3, it gives a cycle length of 45 BV or 22.5 hours. Regeneration levels are 80-100 g HCl/L_R and 60-80 g NaOH/L_R. Of course, these are only indications, the exact requirements for the column designs depend on the engineering companies.

In order to decolorize the syrup, a decolourization unit is placed in front using SBA resins in the Cl⁻ form:

Decolorization→decationization →inversion →deacidification

Depending on the color content of the syrup, one or two columns in series are used for decolorization:

SBA (Cl) $\rightarrow$ SBA(Cl) $\rightarrow$ SAC(H^+) $\rightarrow$ SAC (H^+) $\rightarrow$ WBA

Decolourization can be done at a higher temperature, for example 70-80°C, to decrease viscosity and work at higher velocities and therefore less resins. Then cool down to 40°C for inversion. The decolorizing resins in this flow sheet work under similar principles as those discussed above. The number of columns will depend on the color content of the feed solution.

In all the above discussed ion exchange flowsheets other units are added in order to complete the treatment and produce liquid sugar meeting end-users specifications. These units may include filtrations, ultrafiltrations or adsorbent polishing.

Except using crystalline sugar as raw material, it is conceivable to use fine syrup, that is decolorized syrup from the sugar mill as starting solution (Lancrenon and Paillat, 2005). The fine liquor has a color of 80-200 ICU and ash of about 0.1-0.2%. the flowsheet therefore to consider is:

crystallized sugar $\rightarrow$ remelt $\rightarrow$ decolourization $\rightarrow$ mixed bed demineralization$\rightarrow$GAC

Molasses desugarization

Beet sugar molasses have usually 60% purity while cane sugar molasses have 40% purity. A beet sugar plant loses about 14% of the sugar entering the plant to molasses. With a molasses desugaring by chromatography process it loses 6%. (Asadi, 2007). Chromatography can be used in separating sugar from non-sugars in order to recover sugar from molasses.

Chromatographic separation of the sucrose and non-sucrose from molasses is based on ion exclusion. Ionic species are excluded from the resin due to the Donnan equilibrium while neutral molecules are not (page 28). As a result, if a solution containing sugar and salts is injected into a chromatographic resin column and eluted with water, salts travel quicker down the column while sugar spends more time at the resin beads and comes out later. Betaine interacts even stronger with resins than sugar and comes out after sugar. Betaine is a neutral molecule, more exactly a zwitterion, having a quaternary ammonium group and a carboxylic group, $(CH_3)_3N^+CH_2COO^-$. The two groups form an internal salt so that the net charge of betaine is zero.

Gel type SAC resins are used as the stationary phase. The resin parameters that affect the efficiency of separation (that is, the ability of the column to produce narrow, symmetrical peaks) and the selectivity (the degree of separation of the peaks) are: the resin crosslinking density (DVB level), the resin particle size and particle size distribution and the ionic form. The lower is the crosslinking of the resin, the faster is the diffusion through the resin beads and the better will be the separation efficiency. Similarly, the smaller the particle size of the resin is, the better is the efficiency. The limit however for the particle size is dictated by pressure drop. The smaller the particle size is, the higher is the pressure drop. Also, the limit for the degree of crosslinking is set by the oxidation resistance of the resin, especially in solutions of reducing sugars. A narrow particle size distribution of the resin (along with a uniform packing) ensures a uniform flow distribution and produces narrow peaks. The resin works in the potassium form which in fact is the ionic form in equilibrium with the molasses. If the molasses contain hardness, Ca^{2+}, Mg^{2+}, then the resin is less efficient. This is possibly due to the fact that the resin in the Ca^{2+} form has lower moisture content and diffusion becomes slower. For example, a chromatographic, gel SAC resin in the H^+ form has a moisture content around 57%, the same resin in the Na^+ form has around 51% and in Ca^{2+} form 46%. Therefore, if hardness exceeds 500 ppm CaO the molasses should be decalcified. In cane molasses desugarization where there is more suspended matter than in beet sugar, the use of membranes to remove suspended matter before softening and chromatographic separation has been tested (Kochergin*et al*, 2000). Following is a flow sheet for cane sugar molasses desugarization (Kwok, 2006):

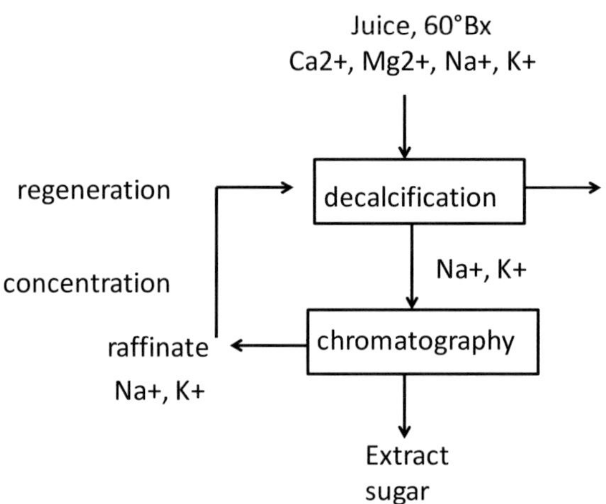

The feed solution is diluted molasses to about 50-70°Bx, temperature 80°C and the eluent is water. When the resin comes in contact with the feed solution, sucrose is distributed about equally between the resin and the solution phase while the ionic species remain to a large extent in the solution phase. During elution with water, the excluded ionic species spend less time inside the resin beads and come out first, followed by the sucrose peak. If amino acids or betaine are present, these come out after sucrose and can be collected separately. The sucrose fraction has typically a purity of 90-95% and can be recycled in the thin juice. The low-purity by-product can be concentrated and used as animal feed. It is possible to obtain a betaine-rich by-product either by pulse method systems or by multistage simulated moving bed systems (see below).

Batch chromatography is used to separate sugar and non-sugars since the years 70's. More recent, Simulated Moving Bed (SMB) systems have been used in molasses desugarization that allow a more efficient separation with less resins and less eluent (pages79-80).

SMB is used to separate two components, like sugar from non-sugars. In order to separate three components, like sugar, non-sugars and betaine, new systems have been developed. One system is the Amalgamated Research Inc. (Ari) Coupled Loop® process where the separation takes place in two steps, the non-sugar plus the sugar are separated from betaine in a first step and then sugar from non-sugars in a second step (Groom*et al,* 2009).Another system for separation of three components, sugar, non-sugar and betaine, is the NS2P/FAST of Novasep (see also page 94) which combines batch process and SMB (Burris *et al*, 2009).

4.Sweeteners

In this chapter it is discussed the processing of monosaccharides such as glucose and fructose, disaccharides such as maltose and hydrogenated products, or polyols, such as sorbitol and mannitol, all these with starch as raw material. In addition, the production of inulin and fructose from chicory roots and the glucose and fructose production from date syrups are mentioned in the end of this chapter.

Starch is a polymer of glucose. It is found in corn, wheat, tapioca, potatoes, rice and other plants. Glucose is manufactured from starch, mainly from corn.

After starch is separated from other constituents of the corn kernels (fibers, germ, gluten) it is liquefied by enzymatic or acid hydrolysis to maltodextrins, which are partially hydrolyzed starch of 5-20 Dextrose Equivalents (DE)[11]. The various hydrolysates of starch are composed of monomers (glucose), dimers (maltose), trimers (maltotriose) and higher oligosaccharides. After conversion, the pH is adjusted to 4.5 and

---

[11] Dextrose equivalent (dextrose = D-glucose) is the relative sweetness of oligosaccharides compared to glucose. For example, 20% DE means 20% as sweet as glucose. It is also an estimate of the percentage of reducing sugars present in the total starch product. It describes the degree of conversion of starch to dextrose.

the liquid is filtered to remove proteins and fats. The proteins level of the starch is at the level of <0.3%. By pH adjustment to about 4.5, proteins precipitate out and removed during the filtration step. After decolorization and refining, maltodextrins are evaporated to 77% DS or dried for commercial uses. Refining of maltodextrins can be done by treatment by ion exchange resins (Deleyn *et al*, 2012). A typical flowsheet would be a SAC-WBA.

The maltodextrins are further hydrolyzed to 20-70 DE glucose syrups, or to >95 DE syrups. Proteins in the glucose syrup are between 0.01 and 0.03%. These proteins can be removed during the ion exchange treatment.
During the conversion of the starch to glucose syrup, temperatures are at 120°C and proteins are solubilised with result that color is formed due to caramelization and to the Maillard reactions.

Glucose syrups with 20-70 DE are used in pastry while >95 DE syrups are used to make crystallized dextrose for food or pharmaceutical applications. Glucose can be isomerized to fructose to make high fructose corn syrups (HFCS) containing 42% fructose (42 HFCS). By chromatographic separation of glucose and fructose, the fructose rich part, containing 90% fructose, can be used to make, by blending with the 42 HFCS, 55 HFCS. By hydrogenation of fructose-rich and glucose-rich syrups, mannitol and sorbitol are obtained respectively.
The HFCS of different compositions are also called in Europe isoglucose.
These are illustrated in the following figure.

146

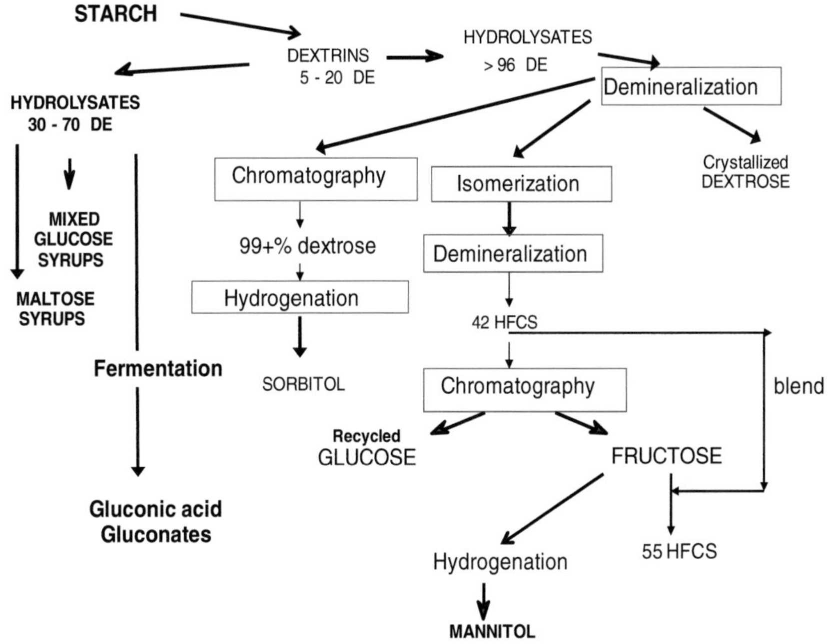

Figure 4.1 Sweeteners from starch

Glucose demineralization

The glucose syrups are first filtered to remove precipitated proteins, oil and other insoluble materials and then decolorized on carbon to remove colour, taste and odour compounds. The syrups are then demineralized over ion exchange resins.

147

The main impurities contained in glucose syrups are inorganic ions (ash), Na^+, Ca^{2+}, Mg^{2+}, Cl^-, SO_4^{2-}, organic acids, peptides and color compounds. Typical levels of ash are 0.5% of the dry solids for corn syrups, 0.3% for wheat and 0.2% for tapioca syrups. Organic acid impurities are in the range of 0.05 to 0.1% of the dry solids.

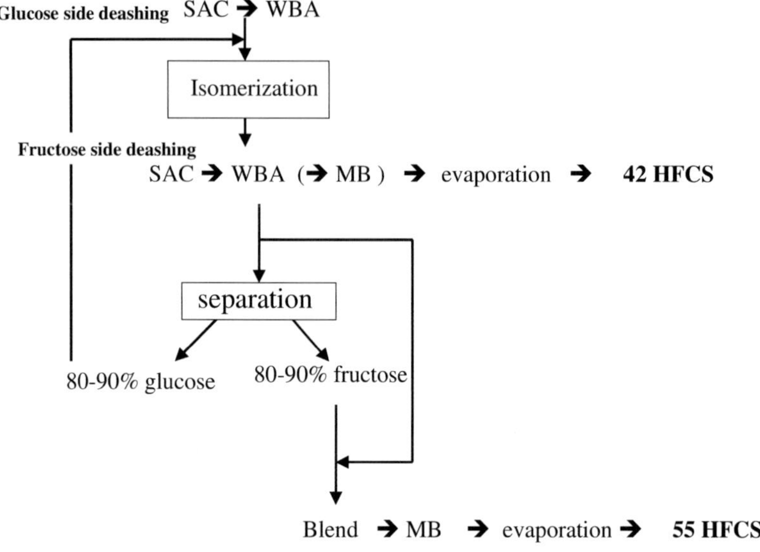

Figure 4.2   Glucose syrups demineralization

Color depends on the pretreatment on carbon before the ion exchange but it can be in the range of 200-300 ICU. Temperature of the syrup is 30-50°C, pH 6-6.5 and concentration before the IER in the range of 35-50 % DS.

148

On the glucose side (see figure 4.2) demineralization is achieved with MR type strong acid cation exchange resins in the H^+ form followed by weak base anion exchange resins in the free base form. The following reactions take place:

$$\mathcal{R}\text{-}H + Na^+Cl^- \rightleftharpoons \mathcal{R}\text{-}Na + H^+Cl^- \qquad (4.1)$$

$$\mathcal{R}\text{-}CH_2N(CH_3)_2 + H_2O \rightleftharpoons \mathcal{R}\text{-}CH_2NH^+(CH_3)_2\,OH^- \qquad (4.2)$$

$$\mathcal{R}\text{-}CH_2NH^+(CH_3)_2OH^- + H^+Cl^- \rightleftharpoons \mathcal{R}\text{-}CH_2NH^+(CH_3)_2Cl^- + H_2O$$
$$(4.3)$$

The SAC resin is of the macroreticular type one reason being physical stability because of the high syrup concentration and the osmotic shock that the resin undergoes at every cycle. It removes inorganic cations, aminoacids and part of color. The MR structure allows also the removal of larger molecular weight peptides and some color bodies. WBA exchange resins remove strong acids generated at the SAC decationization resin, weak acids that are stronger than carbonic acid and color. Strong acid cation exchange resins fix stronger the inorganic cations Na^+, Ca^{2+} and Mg^{2+} than the aminoacids and therefore aminoacids and peptides leak first. WBA exchange resins fix stronger the strong acids, HCl or H_2SO_4, than the organic acids and the color leaked from the cation exchanger and therefore color, aminoacids and organic acids leak ahead of the strong acids. Leakage of aminoacids from the ion exchange unit may cause color formation upon aging due to Maillard reactions. Therefore, the removal of N-containing compounds by the ion exchange unit improves the color stability of the glucose syrup by preventing Maillard reactions from taking place.

The service cycle is monitored at the exit of the WBA resin where it is measured conductivity, pH and color leakage. Conductivity and pH breakthrough at about the same time while colour breaks through first. The cycle stops preferably when colour starts leaking. If at the SAC resin the inorganic cations break through before the WBA breaks through, then conductivity at the exit of the WBA starts rising while pH remains high because WBA resins do not split salts and therefore salts will start leaking from the WBA resin. In order to avoid that the cation exchange resin leaks ahead of the anion, the unit is designed with an excess of cation capacity compared to the anion. Since the total exchange capacity of the SAC resins is in general higher than that of the WBA resins, then, provided that the regeneration level for the cation exchange resin is sufficient, a cation to anion resin volume ratio of one is enough to ensure higher cationic capacity. Nevertheless, having a little more cation resin or a little more anion resin is justified: more cation ensures low leakage of aminoacids and peptides thus protecting the WBA resin from fouling and producing a stable syrup while more anion exchange resin ensures lower color leakage. The choice of the resin ratio depends on the characteristics of the glucose syrup to demineralize. Typical break through curves from the WBA are shown in figure 4.3. The pH of the treated syrup during the loading cycle is about 6-7 while when the acidity starts breaking through, the pH decreases. Usually, the cycle stops when the pH reaches a value of around 4-5 where conductivity is about 20-50 µS/cm.

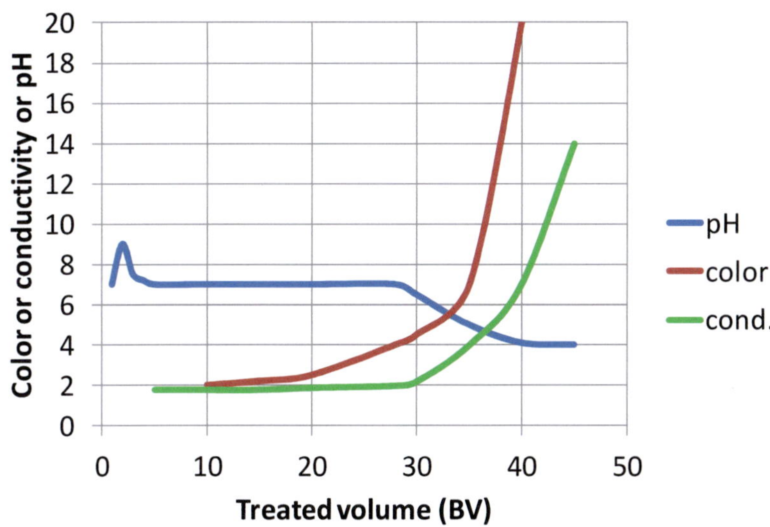

Figure 4.3 Breakthrough curves from WBA resin (arbitrary units)

The resins are regenerated with 80-100 g HCl/L_R using 8% HCl and 60-70 g $NaOH/L_R$ using 4% NaOH. The SAC can be regenerated in co-current or counter-current direction. It is possible to use NH_4OH to regenerate the WBA if this does not have environmental or health problems.

The regeneration sequences are:

1) Sweetening off: demineralized water pushes the syrup out of the resins, going from the SAC to the WBA, until the effluent from the WBA resin is less than 1°Bx. At a specific flow rate of 3 BV/h it should take about 2 BV of water or 40 minutes.

3) Backwash, both cation and anion resins (40 min)

151

4) Chemical injection to both resins. 8% HCl passes through the SAC resin at a specific flow rate of 1 BV/h and 4% NaOH through the WBA resin at 1.5 BV/h. (1 h)
5) Slow rinse: demineralized water pushes the regenerant out of both resins at the same velocity as the regeneration. This ensures the efficient end of the regeneration. 0.5-1.0 BV are sufficient (30-60 min)
6) Fast rinse: 5 BV of water at 5 BV/h through each resin. At the end, the water passes from the cation to the anion resin and conductivity is checked at the exit of the WBA resin. Fast rinse stops when the desired conductivity is reached, usually less than 30 μS/cm (minimum 1 h)
7) Sweetening on: syrup passes through the cation and anion resins until at the exit of the anion resin the syrup concentration reaches >90% of the feed concentration.

The above steps take about 4 to 4.5 hours, provided the fast rinse step takes one hour which often means that the WBA exchange resin is in good condition.

Every about 15 cycles, a clean-up regeneration (cross regeneration) can be performed in order to clean the cation and the anion resin from impurities that do not come out during regeneration. This cross regeneration is performed after a normal regeneration. After cross regeneration, standard regeneration should be performed at a double level for the SAC resin and simple level for the WBA resin. The cross regeneration consists in passing a 4% NaOH solution through the SAC resin to dissolve any proteinaceous material that have accumulated on the resin and a 5% HCl solution through the WBA resin to dissolve organic acids.

With one pair of cation-anion resin the product syrup conductivity is less than 50 µS/cm and decolourization 70-80%. For a better removal of organic acids, peptides and colour, two pairs of cation-anion exchange resins in series can be used, with a third pair on regeneration or stand-by. The end-point is when the primary pair breaks through. When the primary pair is exhausted it goes to regeneration, the secondary moves to the primary position and the freshly regenerated pair goes to the secondary position. With two pairs, conductivity is less than 10 µS/cm and colour less than 30 ICU. In addition, the more complete removal of aminoacids increases the heat stability of the syrup in preventing the formation of color due to Maillard reactions.

Many styrenic WBA resins have a small content,about 5-15% of the total exchange capacity, of strong base groups. These strong base groups in the regenerated, OH^-, form can cause an alkaline isomerization of glucose to fructose (MacLaurin and Green, 1969) as well as degradation resulting in the formation of organic acids (Ellis and Wilson, 2002; Novotny *et al*, 2008) and the appearance of color. A level of strong base groups higher than 12-15 % of the total exchange capacity can cause significant isomerization at 50°C. These groups at the working temperature of 50-60°C degrade progressively by the Hofmann degradation to give weak base groups and therefore their effect lasts only for a short period of time after their installation in the columns.

Another feature of the WBA resins is their relatively high swelling going from the regenerated to the exhausted form. For example, a typical WBA resin swells 20-25% in water going from the regenerated free base form to the Cl^- form. They swell even more with organic acids. For example a styrenic WBA resin going from the regenerated to the lactate form can swell up to 70% when the resin is fully converted to the lactate form. The repeated high swelling-shrinking of these resins during service and regeneration cycles can cause a premature break down causing high pressure drop and resin losses.

The presence of strong base groups can reduce the swelling of WBA resins to a certain extent because the strong base groups swell in the opposite direction than the WBA resins: they swell going from the exhausted form to the regenerated. Therefore, the more the strong base of the WBA resins, the less they swell. However, because of the isomerization of the glucose to fructose due to the strong base groups in the OH^- form mentioned above, with the associated problems, the strong base capacity should be kept to low levels (below 15%).

In order to avoid these problems related with strong base groups, gel type acrylic WBA resins have been tried. Acrylic resins are 100% weak base resins, that is, they do not have strong base functionality. Their gel acrylic matrix gives them a certain elasticity which gives them higher physical stability. At the same time this elasticity may give a higher pressure drop at high fluid velocities and high temperatures. They are alittle stronger bases than styrenic resins. Thus, the pH of the effluent syrup from the WBA resin in the middle of the cycle is 6-7 for styrenic resins while for acrylic is 7-7.5 which should not be of a problem for the syrup quality. In addition, acrylic resins are more hydrophilic and regenerate more efficiently than styrenic resins. The drawback of acrylic resins however is their long rinse requirements after regeneration.

Because of incomplete regeneration, the IER especially the WBA resin become fouled after a certain time of use. Resin pore structure is an important parameter affecting fouling. This fouling causes a decreased operating capacity and for the WBA resins, a long rinse. The fouling matters, colors, peptides, contain carboxylic groups which after regeneration with NaOH are left in the Na^+ form. During rinse, this Na^+ is hydrolyzed and comes out as NaOH resulting into a high pH. Rinse recycling can be practiced in order to save rinse water, this however on one hand consumes cationic capacity and on the other, it may takea prohibitive time. In order to retard as much as possible this fouling from happening, the cationic cycle can be controlled to stop early enough in order to avoid too much proteinaceous material going to the anion resin, perform cross regenerations, especially on the anion resin, and periodically perform a caustic brine treatment on the WBA, consisting in passing a 10% NaCl+2% NaOH solution at warm temperatures and at a slow specific flow rate, giving as much as practical long contact time. It is possible to estimate the degree of fouling by determining the amount of carboxylic groups present in the fouled WBA resin. For this, after a NaOH treatment, the anion exchange groups are converted to the regenerated form while the carboxylic groups will be as Na^+ salts. By passing a given quantity of HCl, the anion exchange groups will fix the HCl while the carboxylic groups will release all Na^+ which will go out with the solution. The total HCl passed through the resin minus the remaining free acid in the effluent will give the anion exchange capacity of the resin while the total chlorides minus the free acid will give the carboxylic content.

If ammonia is used as regenerant for the WBA exchange resin, then the resin requires less rinse water than with NaOH regeneration, the NH_4OH coming out with the rinse water giving lower pH. In addition, NH_4OH does not regenerate the strong base groups so thatwith new resins still having strong base groups the problems related to glucose degradation at high pH are avoided. On the other hand, NH_4OH regenerates less efficiently color compounds fixed by the resin.

For the design of a glucose demineralization plant, a key parameter is the operating capacity of the resin. Because kinetics are rather slow in this case, especially for the anion exchanger, the operating capacity depends on the specific flow rate of the syrup and the concentration of the salts to be removed. Assuming a typical salt concentration for a given syrup, the operating capacity depends then on the specific flow rate.

Consider as an example the demineralization of 10 m3/h corn glucose syrup at 40° Bx. We choose to use a specific flow rate of 3 BV/h and a cation to anion resin ratio of 1. This means that the cation and the anion exchange resin volume will be 10/3 = 3.3 m^3 each. The cycle time is calculated as follows:

The 40°Bx syrup has a density of 1.176. This brings the flow rate to 11.76 tons/h. Since the syrup has 40°Bx, it means that 4.704 tons dry mater per hour pass through the resins. The ash content is 0.5% of the dry mater, therefore, 0.02352 tons ash per hour have to be removed by the resins. Assuming that ash is NaCl, there are 402 equivalents of NaCl to be removed every hour. To this, it is added about 50 equivalents per hour of organic acids, so that the total ions to be removed are approximately 450 eq/h. Taking as resin capacity, both for cation and anion exchanger, 0.85 eq/L_R, this means that 450 / 0.85 = 529 L resin will be needed per hour. Since our design has taken 3.3 m3 of resin, this results to a cycle time of about 6.2 hours. This is in agreement with the regeneration time estimated

156

above to 4-4.5 h. There is in fact an extra 1.5-2 hours where the resins will be on stand-by.

If the resins become fouled, then on one hand the operating capacity drops and on the other the fast rinse time increases. If the operating capacity becomes 0.75 eq/L_R instead of 0.85 then the cycle time goes down to 5.6 hours. Then there is only 1-1.5 extra hours to finish the fast rinse and start the following cycle. If more safety margins are desired, then more resin needs to be installed.

After the demineralization of the glucose syrups some salts are added back to facilitate the enzymatic isomerization of glucose to fructose, where a 42 HFCS (42% fructose) is produced, typically at 60°C and a pH 7-8. These salts need to be removed before evaporation or chromatographic separation of glucose and fructose (see fig.4.2, fructose side demineralization). The same principles apply here as those discussed above. The difference is that the syrup after isomerization has less ash and less color compared to the syrup from the hydrolysis of starch. The syrup on the fructose side has typically ash about 0.2% of DS, color 50-60 ICU and 45°Bx.
One option here is to use a mixed bed (MB) of a SAC and a SBA type 2 resin as a polisher after one pair of SAC-WBA resins. The resins in the MB are of MR type and there is one part of SAC in the H^+ form and two parts of the SBA resin in the OH^- form in the mixture. The loading specific flow rate is 3 BV/h and the temperature 40-45°C. Regeneration is performed with 100 g HCl/L_R using 7-8% HCl solution and 90 g NaOH/L_R using 4% NaOH solution at 20-40°C. Two chains of SAC $\rightarrow$ WBA $\rightarrow$ MB work in parallel, one on loading and the other on regeneration.

Synthetic adsorbents or a mixed bed, as shown in figure 4.2, can be used as polishers at the end of the 42HFCS or the 55HFCS process. Styrene-DVB hypercrosslinked polymer types or formophenolic adsorbents are frequently used in this application.

Glucose-fructose separation

Enzymatic isomerization of glucose to fructose with glucose isomerase produces a mixture containing 42% fructose, 53% glucose and 5% higher saccharides. The mixture is refined with carbon filters and ion exchange demineralization, as discussed above.

The fructose content of the mixture is enriched by chromatographic separation with a SAC resin in the Ca^{2+} form. Fructose, glucose and water form ligand complexes with the Ca^{2+} of the resin. Fructose forms stronger complex than glucose due to more favorable configuration of the OH groups. As the two saccharides pass through the resin bed, fructose is retarded compared to glucose so that glucose comes out first. This is an example of ligand exchange chromatography. The strength of the complex formed depends on the configuration of the OH groups of the saccharide but also on the ionic form of the resin. Depending on the ionic form of the resin, the strength of the complex has the following order:

$$Na^+ < Zn^{2+} < Ca^{2+} < Pb^{2+}$$

With the DVB content of the commercially available resins, the oligosaccharides that may be present in the mixture have too large size to be able to enter and diffuse inside the resin beads. They are excluded by the resin due to their large size and they

158

exit the column first. In this case we have simultaneously ligand exchange and size exclusion chromatography.

On the dextrose side, if after deashing dextrose is hydrogenated to produce sorbitol, then using a SAC resin in the Na^+ or K^+ form, glucose is separated from oligosaccharides by size exclusion chromatography. Glucose enrichment is obtained at better than 99% purity.

In glucose-fructose separation the temperature is in the range of 60-70°C and the syrup 50-60 % DS. Increasing temperature improves the separation but it increases the color. The flow rate of the eluent affects the separation (low flow rate increases separation) but decreases the productivity. Increasing the syrup loading decreases the separation but increases productivity. Under the conditions of glucose-fructose separation, the operating conditions, temperature, pH (it is slightly acidic), flow rate and the presence of fructose favour the formation of HMF which is considered as color precursor. For that reason, the pH control in the process becomes important. In fact, both glucose and fructose form HMF by dehydration but glucose degrades at slower rates giving lower yields compared to fructose (Gomes *et al*, 2015). Here, the isomerisation of glucose to fructose and favors HMF formation.

The main mechanisms of color formation in glucose syrups due to HMF are (Ramchander and Feather, 1974) production of HMF and subsequent condensation with itself or other aldehydes and condensation with aminoacids.

The same characteristics of the resin affect the resolution of the two peaks, as discussed in pages 77-78 and 119, that is the DVB level and the particle size. The lower the DVB level and the smaller the particle size, the better is the resolution. Here, the limit for the DVB level is again the oxidation stability of the resin. The presence of reducing sugars in combination with the dissolved air (oxygen) in the feed makes the resin more liable to be oxidized. If the resin has low degree of crosslinking, then oxidation of the resin leads in a short time to an increase of the moisture content accompanied by an increase of pressure drop. Resins with higher degree of crosslinking also oxidize by dissolved O_2 but the effect is seen after a longer time. In any case, it is recommended to de-aerate the feed solution prior to the chromatographic resins in order to extend the life time of the resin.

In a SMB separator, the 42% fructose syrup enters at the feed port. As the syrup moves in the direction of the eluent flow, fructose moves slower than glucose. This results in the formation of a glucose-rich band, a mixed glucose and fructose band and a fructose-rich band. Eventually, the glucose band reaches the raffinate port and comes out. When the mixed band reaches the raffinate port, then the columns of the separator move one position upstream, or the entrance and exit ports move one position in the direction of the eluent. As the separation continues, the fructose band separates from the mixed band and eventually reaches the extract port where it comes out.

Typically, the raffinate (glucose) returns to the isomerisation and the fructose-rich product (the extract) is blended with the 42% fructose syrup to make 55 HFCS.

In order to have good separation, a number of operating parameters should be controlled carefully. It is very important to keep flow rate constant to ensure that the bands are found in the right position at the right time. The amount of resin should be exactly the same in each column and the resin characteristics should be the same in each column. This is important to take into account because there are batch-to-batch variations in the

160

resin characteristics. Therefore, when filling an installation with resin, either the total resin quantity must first be mixed together or else, they have very tight specifications for the different batches of resin.

Sugar alcohols

Hydrogenated saccharides are called polyols or sugar alcohols. Hydrogenated starch hydrolysates (HSH) are produced by hydrogenation of starch hydrolysates of various DE, so in fact they constitute a family of products. Hydrogenation of glucose gives sorbitol (also known as glucitol) and hydrogenation of fructose gives a mixture of the isomers sorbitol and mannitol. Maltitol is hydrogenated maltose which is a dimer of glucose. Hydrogenation of the pentose xylose gives xylitol.

Hydrogenation is performed in the presence of a catalyst at specified pressure and temperature. Catalysts include Raney Nickel, supported nickel and other metals. After hydrogenation, the solution is decanted and syrups are filtered off. Depending on the catalyst used, if catalyst is found dissolved in the syrup it must be removed and ion exchange is a suitable means.

Polyols can be purified by ion exchange demineralization. The solutions pass through a SAC in the H^+ form followed by an anion exchanger. A SBA resin can be used in the OH^- form or a WBA resin in the free base form or both. Using both, SBA and WBA resins, has advantages because the feed solutions contain weak and strong acids (after the SAC resin). SBA of type 1 or type 2 can be used. The two anion exchange resins, weak and strong, can be used in the same column. For a high quality syrup, a double pass can be used and/or with a polishing mixed bed with a SAC/SBA resins. If nickel is also present in the feed solution, a WAC resin in the H form can be used before the SAC resin to remove and recover nickel, or food grade iminodiacetic type resins which are very selective for nickel.

Glucose is hydrogenated to sorbitol generally at a temperature of 120-160°C and hydrogen pressure of 70-140 atm with a supported nickel or Raney nickel catalyst. As mentioned above, glucose can be enriched using size exclusion chromatography to better than 99% glucose by using a SAC resin in the Na^+ or K^+ form.

The feed solution passes through the resins at a specific flow rate of 2 BV/h at 35-50° C. Depending on the feed composition, the operating capacity of the resin is 0.3-0.4 eq/L_R. The low capacity may be due to the difficult removal of gluconic acid.

Alkali-stable sorbitol is required in applications such as toothpastes, especially sodium carbonate toothpastes, where sorbitol syrups are used as humectants, in order to prevent color formation with time of storage. The color formation is due to the alkaline degradation of reducing sugars. In order to achieve color stability of sorbitol, the reducing sugars are first degraded at alkaline pH followed by purification on IX. The alkaline degradation can be done at the same reactor as the hydrogenation (Salom et al, 2002) while the purification consists of passing the syrup over a SAC resin followed by a SBA resin. Specific flow rate is 1-2 BV/h and temperature 20°C. Another way to degrade the reducing sugars is with a SBA resin in the OH- form at elevated temperatures and specific flow rates below 1 BV/h (van Lancker, 2007).

Hydrogenation of fructose gives a mixture of mannitol and sorbitol at proportions that depend on the hydrogenation conditions.

Mannitol and sorbitol can be separated by ligand exchange chromatography using a SAC resin in the Ca^{2+} form (Melaja and Hamalainen, 1975; Wolfgang *et al*, 1997).

Oligosaccharides

Prebiotics are a range of not digestible, carbohydrate structures, food ingredients that are beneficial to the host by selectively stimulating the activity of a number of bacteria in the colon. One category of prebiotics is oligosaccharides such as galacto-oligosaccharides, xylo- oligosaccharides, fructo-oligosaccharides or polydextrines. Food-grade oligosaccharides are manufactured either by extraction from the plants, or by enzymatic controlled degradation of polysaccharides or enzymatic synthesis from simple saccharides. A general enzymatic manufacturing process includes, after the batch reaction, a decolorization and demineralization by activated carbon and ion exchange. Following the purification, there is a chromatographic separation step to separate mono- and disaccharides from oligosaccharides, using a SAC exchange resin in the Na$^+$ form.

Human Milk Oligosaccharides

Human Milk Oligosaccharides (HMO) is a family of diverse oligosaccharides abundant in human milk, playing specific functions with potential benefits for the infant. Almost 200 different species of oligosaccharides have been identified in human milk. They consist of glucans of less than about 25 monosaccharides of complex structures. Because of this complexity, their synthetic production is challenging. They can be produced by chemical synthesis, enzymatic synthesis or by fermentation. Ion exchange has been used in the purification of HMO. From fermentation broths, after filtration to separate the biomass from the broth, strong acid cation exchange resin followed by strong base anion exchanger in the formiate form have been used to remove ionic substances (Jennewein, 2015) or SAC-WBA exchange resins (Chassagne *et al,* 2017).

Inulin

Inulin is a polymer of fructose with glucose molecules as the terminal units, found in chicory roots and some other plants like topinambur, Jerusalem artichokes or dahlia. The degree of polymerization of inulin varies in the range of 11 to about 60. Fructooligosaccharides (FOS) have a degree of polymerization of 3 to 10 and can be obtained by partial hydrolysis of inulin. Inulin and FOS are increasingly used in processed food for their important functional properties. Inulin can be used to replace sugar, fat and flour. It also increases calcium and possibly magnesium absorption and promotes the growth of intestinal bacteria. Upon acid or enzymatic hydrolysis inulin gives a fructose syrup having a fructose to glucose ratio of around 85/15, depending on the raw material. Fructose production from inulin hydrolysis is therefore an alternative way to that from starch hydrolysates, obtained by isomerisation of glucose followed by glucose-fructose separation. Industrial production of inulin is practiced in Belgium and Holland. The major inulin processors are Cosucra Groupe Warcoing and ORAFTI in Belgium and Sensus in Holland. It should be noted that since the European sugar regulations of June 30, 2006, the above mentioned European inulin producers produce only inulin and not fructose any more.

Inulin is extracted from the chicory roots in a similar way as sugar is extracted from beets. Chicory roots are first cut into pieces and inulin is extracted with hot water.The obtained juice is then limed and carbonated where the formed $CaCO_3$ precipitate entrains peptides, colloids and color. By filtration, a raw juice is obtained which can be purified on ion exchange resins and activated carbon by demineralization and decolorization (Franck and De Leenheer, 2005).
Demineralization is achieved with SAC in the H^+ form followed by WBA resin in the free base form. Since the pH at the SAC resin is acidic, low temperatures are used in order to avoid inulin degradation. In order to minimize environmental contamination, H_2SO_4 and NH_4OH are used as regenerants of the SAC and the WBA resins. The effluents are concentrated where K_2SO_4 and

$(NH_4)_2SO_4$ are precipitated out, recovered and sold as fertilizers. The mother liquor is further concentrated and sold as feed.

To produce fructose, inulin is hydrolyzed either enzymatically or with an acid such as HCl. Acid hydrolysis is obtained by controlling three parameters, acid concentration, time and temperature (Nasab *et al*, 2009) in order to optimize hydrolysis and minimize the fructose degradation to HMF (Yi *et al*, 2011). Following acid hydrolysis, the solution is neutralizedwith CaO and treated on ion exchange resins to demineralise and decolorize.In acid hydrolysis of inulin to produce fructose syrups, some HMF formation cannot be avoided which subsequently can give color and off taste problems. Enzymatic hydrolysis avoids these problems provided that pH remains at 4.5 and temperature at 60°C (Franck and De Leenheer, 2005).

Sweeteners from various plants

A large number of natural sweeteners has been identified as plant constituents. These highly sweet compounds fall mainly in the terpenoid, flavonoid and protein categories. Among these sweeteners, steviol glucosides have obtained the more widespread interest.
Stevia is the name of a family of plants of which the species *stevia rebaudiana* commonly known as simply stevia, is widely grown for its sweet leaves. This sweet taste is due to steviol glucosides terpenes. The two primary compounds found in the leaves of stevia are stevioside and rebaudioside A. They have about 300 times the sweetness of sugar and have a negligible

166

effect on the glucose level in the blood, so they are attractive as natural sweeteners for people with carbohydrate-controlled diet. However, the consumption of these steviosides is limited due to the after taste bitterness, astringency and grassy taste due to the presence of compounds such as diterpene/alkaloids present in the leaves. Other compounds found in stevia leaves include triterpenes, sterols, flavonoids, volatile oil constituents, gums, pigments and inorganic substances.

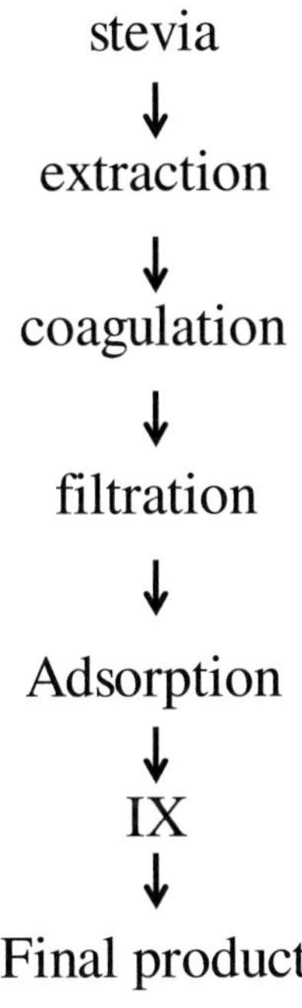

Figure 4.4   Flowsheet for stevia processing.

The extraction of steviol glucosides from the leaves of stevia is achieved with warm or hot water either in batch or continuous reverse flow systems (fig. 4.4). It follows a filtration and a coagulation step using $Ca(OH)_2$ to remove pigments, phenolic compounds and proteins. The precipitate is removed with filtration or centrifugation and the filtrate is treated first with a polymeric adsorbent. As adsorbent, styrenic-DVB or acrylic types have been used, at 2-4 BV/h and ambient temperature. The adsorbent adsorbs the steviol glucosides and the spent extract comes out. Elution of steviol glucosides from the adsorbent is done with about 2 BV of an alcohol, ethanol in general. The eluate that contains the steviol glucosides along with some pigments and salts, is then treated with activated carbon and filtered. Following this, the liquor is treated on ion exchange resins using a gel type SAC followed by WBA resins (Giovanetto, 1990; Jonnala *et al*, 2006) or a macroporous SBA resin. Flow rate 2-4 BV/h. Before the SAC-WBA resins, a high moisture gel type SBA resin in the Cl^- form has also been tried to decolorize the juice (Payzant *et al*, 1999).

The purified alcoholic extract is concentrated with NF, which retains steviol glucosides and allows the alcohol to pass through and recycled. The concentrated liquor is further concentrated by evaporation, treated again over activated carbon and finally is spray dried.

An alternative flowsheet is to place the adsorbent after the IX purification. The adsorbent adsorbs steviol glucosides from where the glucosides are eluted wirh alcohol, followed by crystallization.

Date sugar

The sugars contained in dates from the Middle East cultivars are mainly invert sugar (glucose and fructose) and low levels of sucrose while in some varieties in North Africa, about one-half is sucrose. After extraction of the sugars at 60-70°C, the juice is filtered and decolorized on activated carbon before being demineralised on ion exchange resins. An UF step may or may not follow the filtration step. Following demineralization and concentration, the syrup can be used as feedstock to produce high fructose syrups. For example (Al Eid, 2006), 40 % DS date syrup was used to separate fructose from the rest of the sugars, using a SAC chromatographic grade resin in the Ca^{2+} form. The separated fructose can then be used to produce 55 HFS. It is conceivable that date syrups can be used as raw material to produce sorbitol and mannitol from the separated glucose and fructose, citric acid, or polyols in a similar way as from corn syrups.

5.Milk whey

Whey is the liquid part of the milk that remains after cheese has been coagulated and removed. It contains compounds such as proteins, lactose and minerals. Products that come from milk whey include *demineralized whey powder, lactose, proteins and mineral products rich in calcium and phosphates.* These products find uses in sports drinks, infant formula preparations, bakery, confectionery and yoghurts.

Cheese whey contains minerals originating from the milk and from salts added during the cheese process. A typical composition of milk whey is given in table 1.

Table 1: composition of cheese whey (weight %)

Fat	0.05
Lactose	4-5
Casein protein	0.20
Whey protein	0.6-0.65
Minerals	0.5-0.7
Lactic acid	up to 0.8
Citric acid	0.1
pH	5 (acid whey) or 6.3 (sweet whey)
dry matter	6-6.5

The level of the minerals in the whey makes it unsuitable for human or animal consumption. A typical demineralization flowsheet is shown in figure 5.1

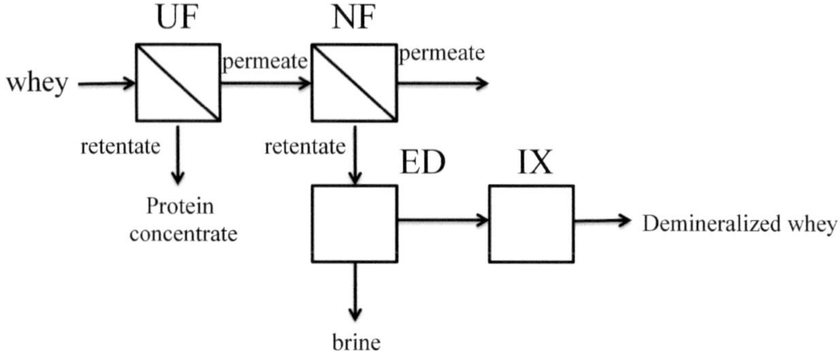

Figure 5.1

In some flowsheets there is no ultrafiltration (UF) unit in front. UF however except that allows the recovery of protein concentrates, it reduces the fouling of the NF unit. In the flowsheet of fig. 5.1, nanofiltration (NF) produces about 30% demineralised whey, electrodialysis (ED) produces 50-70% demineralization and then ion exchange (IX) achieves better than 90% minerals removal, this final demineralised whey being used among others in infant food fotmulas.

The NF membranes work properly up to about 20% solids, above which, precipitation of calcium phosphates prevent further demineralization. One way to improve the demineralization of the NF unit is to decalcify the solution by a SAC resin in the Na^+ form. Here, the presence of calcium-

172

complexing molecules like citrates, lactates or phosphates, limit the softening power of the SAC resin, as already discussed on pages 33 and 34. This can be faced by using a SBA exchange resin in the Cl⁻ form before the softening resin (Theoleyre and Gula, 2004) which removes the complexing molecules thus resulting into a more efficient softening (figure 5.2).

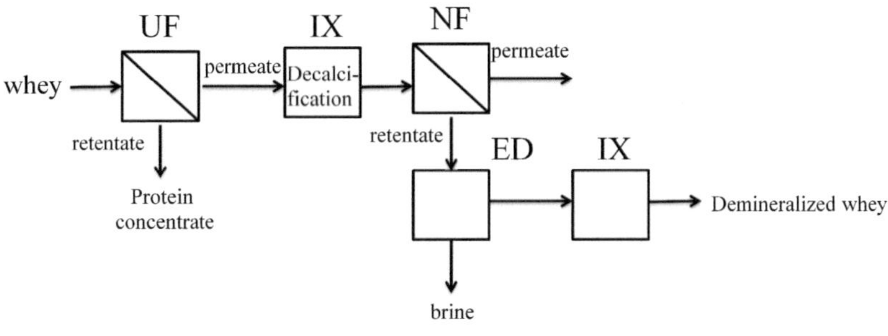

Figure 5.2

Demineralization by ion exchange is achieved by treating the whey through cation exchange resins in the H^+ form followed by WBA exchange resins. Typically, ions to be removed by ion exchange are Ca^{2+}, Na^+, K^+, Mg^{+2}, Cl^-, PO_4^{3-}, citrates, lactates. Their approximate concentration in whey is as follows, in grams per liter (de Wit, 2001): Ca^{2+} 0.6; Mg^{2+} 0.1; K^+ 1.5; Na^+ 0.5; P 0.7; Cl^- 1.1.

The resin types to use are gel-type strong acid cation exchanger in the H^+ form followed by weak base anion exchanger in the free base form. Styrenic, MR type resin can be used as WBA but a gel type acrylic resin can be envisaged. The advantage of the acrylic gel type WBA is that it gives less protein losses than the MR type styrenic. The gel structure of the acrylic WBA resins does not allow large size proteins to enter the resin beads while

173

the macropores of a MR type styrenic resin can trap large molecules in the pores. On the other hand, the drawback of acrylic weak base anion exchange resins is the long rinse requirements.

In order to avoid bacteriological growth, temperatures below 10°C are used.

The presence of citric and lactic acids at high concentration may cause physical stability problems on the WBA resin because of the significant swelling of the resin in those acids. The same remarks apply here as those discussed in glucose deashing, page 153-154.

The reactions of the whey demineralization are similar to those shown in page 149, reactions (4.1) to (4.3).

The whey to be demineralized enters the SAC resin in the H^+ form where the cations are exchanged for the H^+ of the resin dropping the pH of the whey to about 1.5. The acidified solution enters then the WBA resin where the acid solution is neutralized by the anion exchange resin increasing the pH.

Phosphoric acid is a triprotic acid with the dissociation constants indicated below:

$$H_3PO_{4(s)} + H_2O_{(l)} \rightleftharpoons H_3O^+_{(aq)} + H_2PO_4^-{}_{(aq)} \qquad K_{a1} = 7.25 \times 10^{-3}$$

$$H_2PO_4^-{}_{(aq)} + H_2O_{(l)} \rightleftharpoons H_3O^+_{(aq)} + HPO_4^{2-}{}_{(aq)} \qquad K_{a2} = 6.31 \times 10^{-8}$$

$$HPO_4^{2-}{}_{(aq)} + H_2O_{(l)} \rightleftharpoons H_3O^+_{(aq)} + PO_4^{3-}{}_{(aq)} \qquad K_{a3} = 3.98 \times 10^{-13}$$

Since the last two constants are very small, phosphoric acid behaves as monoprotic acid. Still, it is a moderately weak acid. As a consequence, the affinity of the WBA resin is lower for the dihydrogen phosphate (a weak acid) than for the strong H^+Cl^- and in fact $H_2PO_4^-$ leaks first, before Cl^-. The end of the cycle is based therefore on phosphate breakthrough. In comparison, the pKa1 of citric acid is 3.14 and the pKa of lactic acid is 3.60, (the pKa1 of H_3PO_4 is 2.14) therefore phosphoric acid is fixed stronger by the WBA than citric or lactic acid. Consequently, the calculation of the resin volumes for an installation of whey demineralization, it is taken into account the cations (Na^+, K^+, Ca^{2+}, Mg^{2+}) and the anions (Cl^-, $H_2PO_4^-$).

Regeneration is performed with 80 g HCl/L_R for the SAC in counter-current direction or 100 g HCl/L_R in co-current direction and 70 g $NaOH/L_R$ for the WBA resin. The NaOH level for the WBA resin should take into account that the $H_2PO_4^-$ ions fixed on the resin will consume three equivalents of NaOH. The regeneration steps are similar to those in glucose deashing, pages 151-152: sweetening off, backwash, chemicals injection, slow rinse, fast rinse, sweetening on or stand-by. Regeneration takes about 4 hours provided that the WBA resin rinses down in about one hour.

Taking 1.05 g/ml as the density of the whey, 0.5 wt % of minerals means 5.25 g/L. Taking approximately that minerals are NaCl (Eq Wt 58.5), the minerals concentration in the whey becomes about 90 meq/L. Assuming an operating capacity of 1 eq/L_R for both the cation and the anion resin, it follows that the cycle length would be 1000/90= 11.1 BV. Taking the cycle time as 4 hours, the specific flow rate becomes 11.1/4= 2.8 BV/h. For example, if the flow rate of the whey is 10 m3/h then the resin required is 10/2.8 = 3.6 m^3 of SAC and 3.6 m^3 of WBA resin. The loading cycle will be 4 hours and the regeneration time should also be 4 hours.

If the minerals are more concentrated, for example 0.7% by weight, then in terms of equivalents per liter this concentration becomes 125 meq/L. With the same operating capacity as before, that is 1 eq/L_R, the resin will be able to treat 1000/125= 8

BV. With 4 hours cycle, the specific flow rate is 8/4=2.0 BV/h and the resin requirements for the 10 m3/h whey are 10/2.0=5.0 m^3 SAC and 5.0 m^3 WBA.

These are conservative numbers because regeneration time of 4 hours is when the resins are new and in good condition. Also, the feed solution contains in addition lactic and/or citric acids. As mentioned above, there is a partial demineralization before the ion exchange either by a nanofiltration membrane and/or electrodialysis. Monovalent cations Na^+ and K^+ are transferred preferentially through the membranes of ED and therefore there is proportionally more Ca^{2+} and Mg^{2+} in the whey going to the IER. The lower level of minerals allows a more safe design for the ion exchange demineralization which can then remove better than 90% of the minerals and free acids.

Lactose, a disaccharide of galactose and glucose, is the main constituent of cheese whey and is recovered from the UF permeate by crystallization. Lactic acid is produced by fermentation of lactose. Galactose as a monomer is not found freely in nature and therefore production of galactose is obtained from hydrolysates containing galactose. Such a hydrolysate is obtained from hydrolysis of lactose and the separation of galactose from glucose can be achieved with a SAC resin in the Ca^{2+} form (Saari*et al*, 2010) by ligand exchange chromatography in a similar way as glucose-fructose (see page 158) or mannitol-sorbitol separation (see page 163). Galactose is not as sweet as glucose or fructose and it produces a lot of energy with less product. It can be used as sweetener in bakeries. It is also a raw material in the pharmaceutical industry.

Based on ion exclusion chromatography, as was the case with molasses desugarization (page 140), lactose can be separated from mineral salts and proteins. Lactose, a neutral molecule, is distributed equally in and out of the resin beads while mineral salts are excluded and spend only a short time inside the resin. Proteins also do not enter inside the resin due to their size (size exclusion). Upon elution, proteins and minerals come out first and then lactose follows. A SAC resin is used in the Na^+ form (Harju, 2007) with water as eluent.

Removal of lactose from skim milk can be achieved with ion exclusion and size exclusion chromatography, as above. Lactose is separated from proteins and salts of skim milk thus producing lactose-free milk (Harju, 2007) for people that have difficulty in digesting lactose.

Tagatose, an epimer of fructose, is a sweetener that brings fewer calories than the other sugars and also upon digestion, it results in a lower glucose level in the blood. It can be obtained by isomerisation of galactose at alkaline pH using calcium hydroxide as complexing agent.The manufacturing of tagatose is done in two steps (Beadle *et al*, 1992). In the first, lactose is hydrolyzed enzymatically or under acidic conditions to produce a mixture of D-glucose and D-galactose. These two sugars are then separated by ligand-exchange chromatography using a SAC resin in the Ca^{2+} form. In the second step, D-galactoseis isomerized under alkaline conditions in the presence of $Ca(OH)_2$ to form an insoluble complex of $Ca(OH)_2$- D-tagatose. After filtration, the precipitate is acidified with H_2SO_4 to produce soluble D-tagatose and $CaSO_4$. D-tagatose is recovered by filtration to remove the CaSO4 precipitate and the filtrate is demineralised with a SAC in the H^+ form followed by a WBA resin in the free base form. The SAC is a gel-type resin while the WBA resin can be acrylic or styrenic type. The demineralised solution is then concentrated to recover D-tagatose.

Whey proteins are marketed as dietary supplements and commercially come in different forms: whey protein concentrates (WPC) which are products with low fat but high lactose content, whey protein hydrolysates (WPH) which are partially hydrolyzed, easier to digest, whey protein isolates (WPI) which have low levels of fat and lactose. Separation of whey proteins can be achieved by various chromatographic techniques. Gel chromatography allows the separation of proteins from other constituents based on molecular size. Ion exchange chromatography is based on the electrical charge of the whey proteins and is achieved with ion exchange. It allows the isolation of specific whey proteins. This can be achieved by passing whey at a pH about 3, where proteins are positively charged, through a strong acid cation exchange resin. Elution is achieved either by one buffer eluent where all whey proteins are eluted together to obtain whey proteins isolates, WPI, or two buffer eluents to separate alpha-lactalbumin (ALA) from the ALA-depleted WPI, or four buffer eluents to make ALA, ALA-depleted WPI, lactoperoxidase (LP) and lactoferrin (LF) (Dultani*et al*, 2004). A different technique to separate LP and LF from the rest of the whey proteins is to pass whey at the normal pH, 6.5 of sweet whey for example, through a cation exchange resin in the Na^+ form (de Wit, 2001). At this pH, LP and LF are positively charged while the rest of the whey proteins are negatively charged. LP and LF are then eluted with salt solutions of different concentration.

6. Fruit juices and beverages

Apple juice

Apple juice is manufactured by first crushing the apple fruits to form a pulp, called pomace, followed by the addition of enzymes to break down the walls of the cells in order to increase the extraction of the juice. The obtained cloudy juice is centrifuged and then filtered. Before filtration, the juice can be treated with enzymes which break down turbid particles. The juice is then pasteurized for packaging or it is treated further by dehydration to produce concentrates. Clarified concentrates have 70-71°Bx. It exists also an unclarified, cloudy, concentrated juice at 45°Bx.

One major use of ion exchange and adsorbents in apple juice processing is in the production of apple syrups. Apple syrups are solutions that contain essentially sugar and which are obtained from juices after removing minerals and colors. They are used as canning syrups or as additives in soft drinks to give some apple flavor.

The decolorization and demineralization of apple juices can be achieved with the following layout:

$$SAC \rightarrow WBA \rightarrow adsorbent$$

The SAC resin in the H^+ form removes the cations such as K^+, Ca^{2+} or Mg^{2+}, the WBA resin removes the acidity and part of the color and the adsorbent removes the rest of the color and color precursors. Apple juices with a Brix of around 15° contain about 50 meq/L of free acidity and another 40 meq/L of salts. This implies that the cations to be removed are about 40 meq/L while the anions are 90 meq/L. Taking an operating capacity of the SAC resin of 1 eq/L_R and an operating capacity of the WBA resin of 1.3 eq/L_R then the ratio of the WBA to the SAC resin volume should be 1.7 to one. The specific flow rate is 2 BV/h where the BV is based on the anion exchange resin volume. The specific flow rate for the SAC is therefore 3.5-4 BV/h depending on the cation to anion resin volume ratio. Acrylic resins can be used as WBA because of their higher operating capacity and the easier regeneration for the color bodies absorbed.

The adsorbent is typically of styrene-DVB type. Its function is to remove color left from the WBA resin. If an acrylic WBA resin is used, then the combination of acrylic (WBA) and styrenic (adsorbent) structures result into a more efficient decolorization. In the above flow sheet, the adsorbent volume is the same as the SAC volume and therefore the specific flow rate is 3.5-4 BV/h. Regeneration of the adsorbent is performed here with NaOH at elevated temperatures, followed by an acid wash to lower the pH before starting the loading cycle. As acid HCl, H_2SO_4 or H_3PO_4 can be used.

The regeneration sequences of the styrene-DVB type adsorbents are similar to those described in pages 76-77.

Because of the high free acidity of the apple juices, another flow sheet that can be used is (Norman, 1999):

$$\text{Adsorbent} \rightarrow \text{WBA} \rightarrow \text{SAC} \rightarrow \text{WBA}$$

In this flow sheet, the first WBA removes the free acidity, the SAC in the H^+ form removes cations and the second WBA removes the acidity associated with the cations plus any color left. The adsorbent removes color bodies and color precursors and it can be placed in front of the chain because the acidic pH favors the color removal. Placing the adsorbent unit at the end of the chain on the other hand has some advantages to consider because the WBA resin removes part of the color protecting the (expensive) adsorbent thus increasing its life time.

Patulin is a mycotoxine produced by a number of molds in fruit spoilage and it may be a carcinogenic compound.

Patuline structure

Several methods are used to limit the levels of patulin in apples juices like charcoal as well as various clarification processes. Many adsorbents have been tried like ultrafine activated carbon bound togranular quartz to produce a fixed bed adsorbent which removed patulin from apple juice (Cousin*et al*, 2005).

Synthetic polymeric adsorbents have also tried with variable success. One synthetic adsorbent reported to give satisfactory results is a hypercrosslinked adsorbent bearing weak base functionality and having mainly micropores (<20 Å) and a high specific surface area, >900 m^2/g (Lyndon and Miller, 2001). The pore structure was an important factor for patulin removal. Adsorbents with large pores adsorb also other molecules of larger size, like for example color bodies, which compete with patulin for the adsorption sites thus decreasing the capacity of the adsorbent for patulin. In addition, having mainly small size pores (< 20 Å) prevents the large size color compounds from being adsorbed and therefore, there is no, or little, decolorization of the juice which is an advantage if it is not desired to decrease the color of the apple juice.

The adsorbent was regenerated with ammonia. Apparently, NaOH regeneration was not efficient regenerant. The loading specific flow rate was 5-10 BV/h. As an example, patulin was reduced from about 100 µg/L down to 13 µg/L.

Haze formation in apple juice with time is mainly due to the interaction between proteins containing proline and polyphenols (Siebert, 1999). The interaction is non-ionic and takes place between the proline groups of the haze active (HA) protein, and the phenolic group having at least two adjacent OH groups and preferably three. The more proline groups are in the HA proteins, the more haze is formed. Polyphenols are more haze active than monomeric phenols. The polyphenol molecules must have at least two sites that bind with the protein molecule. In this way, polyphenols cross-link the proteins and form large size network structure that cause haze. The maximum amount of haze is obtained when the polyphenol and the proteins binding sites are present in about equal parts. If proteins are in excess or if polyphenols are in excess, like in apple juice, the haze formation is reduced.

The stabilization of apple juice to retard haze formation is obtained by removing one or both of the HA components, proteins or polyphenols. Polyvinylpolypyrrolidone (PVPP) interacts with polyphenols in a similar manner as proline-rich proteins do and is used as stabilizer.

Synthetic adsorbents can also be used to remove polyphenolic material from apple juice thus stabilizing it from haze formation. Styrene-DVB types or phenolic type adsorbent can be used. In the later case, it should be taken into account that phenolic adsorbents contain OH groups which act like weak acid groups capable of exchanging cations for the H^+ of phenol at pH neutral or alkaline. At the acidic pH of the apple juice, the OH groups are too weak acids to be dissociated and therefore they are unable to exchange cations.

In addition to stabilizing the apple juice, synthetic adsorbents will (partially) decolorize at the same time. Regeneration of these adsorbents can be done as described in pages 76-77.

Orange juice

After harvesting and grading oranges the juice is extracted from the fruits and filtered on a stainless steel filter to remove pulp and seeds. From this it is made either a frozen concentrated juice (FCOJ) or a not-from-concentrated (NFC) juice. Juices ready to serve (RTS) are made from concentrates (65°Bx) by blending with water to 12-15° Brix, pasteurized and packaged.

IER have been used in orange juice processing in two ways, in reducing acidity and in removing bitterness.

Orange juice deacidification is performed using a WBA resin in the free base form. The juice is first filtered or centrifuged to separate pulp from the rest of the juice and then the pulp-free part is treated over a WBA resin where the free acidity, consisted mainly of citric acid and some malic,ascorbic and folic acid, is fixed on the resin. As a WBA resin, acrylic type may be preferred here due to its higher physical stability compared to styrenic resins. High physical stability is needed here because the resin swells considerably going from the regenerated form to the exhausted, citrate, form so that the resin undergoes significant swelling and shrinking during the service cycle and the regeneration cycle. In addition, acrylic resins swell somewhat less than styrenic resins.

A WBA resin has a total exchange capacity, depending on the commercial product, from 1.2 to 1.6 eq/L_R. The free acidity of orange juices varies but it can be from 10 g/L as citric acid up to as high as 25 g/L in very acidic juices. As the clarified juice passes through the regenerated WBA resin bed, all free acidity will be fixed by the resin while the effluent juice will have an alkaline to neutral pH. In terms of equivalents per liter the free acidity of the juice would be 150-390 meq/L (taking citric acid as trivalent at the pH of the regenerated resin). A typical specific flow rate is about 5 BV/h. Taking an operating capacity of the WBA resin of 1.2 eq/L_R then within an hour or so the functional groups of the resin will be occupied by citric acid. As the loading cycle continues, the pH inside the resin beads will decrease and citric acid will switch to divalent and monovalent form liberating exchange sites of the resin which can then fix more citric acid. The effluent pH will decrease accordingly and the weaker acids like ascorbic and folic acid will leak through. A pH below 4.6 indicates that the ascorbic and folic acids have been displaced from the resin. After about 3-5 hours the resin will be exhausted with citric acid and the effluent pH will have the same value as the influent. The resin will then be ready to be regenerated. Regeneration of the resin is performed with 4% NaOH at a level of 70 g $NaOH/L_R$. The regeneration steps are similar to those discussed earlier (pages151-152).

At the early part of the loading cycle, the effluent pH will be well above 4.6 where microbial growth starts especially if the WBA resin contains some strong base functionality which means that it can even split some salts and raise the pH to alkaline. In order to avoid that pH remains high with the risk of microbial growth, the WBA resin may be preconditioned after the NaOH regeneration by a dilute solution of citric acid (Chung*et al*, 2007).   This consumes the resin capacity to a certain extent, about 20-30%, but on the other hand it assures an effluent pH no greater than 4.6. A fast flow rate at the beginning of the cycle, 10-15 BV/h, helps keeping the effluent pH to low levels.

Another way to prevent the effluent pH from being too high is to treat only part of the juice and blend the treated with the untreated juice immediately after the deacidification in order to restore the pH to 4.6 and thus to prevent microbial growth (Lineback *et al*, 2006). After blending, the pulp can also be put back into the clarified juice otherwise the juice can be sold as reduced pulp juice.

Bitterness in orange juice is due mainly to limonin, found in seeds but in certain varieties like the navel oranges where there are no seeds, it is found in the fruit. Limonin is an esterification product of limonoic acid d-ring lactone which is not bitter but it is converted to limonin in acidic pH and elevated temperatures. Therefore, in processing navel oranges to make juice, a bitter taste develops. Bitterness in grapefruit juice is due to a flavonoid, the naringin.

Many debittering techniques have been developped including enzyme treatment and adsorbents.

Synthetic adsorbents of styrene-DVB types have been used successfully in orange juice debittering (Norman, 1999; Shaw*et al*, 2000). The orange juice is first treated to separate pulp from the serum, otherwise the adsorbent column becomes clogged quickly. Various techniques have been used to achieve that, such as centrifuge or ultrafiltration membranes (Milnes and Agmon, 1995) promoted by Koch Membrane Systems. The removed pulp is put back into the debittered juice thus reconstituting the original juice.

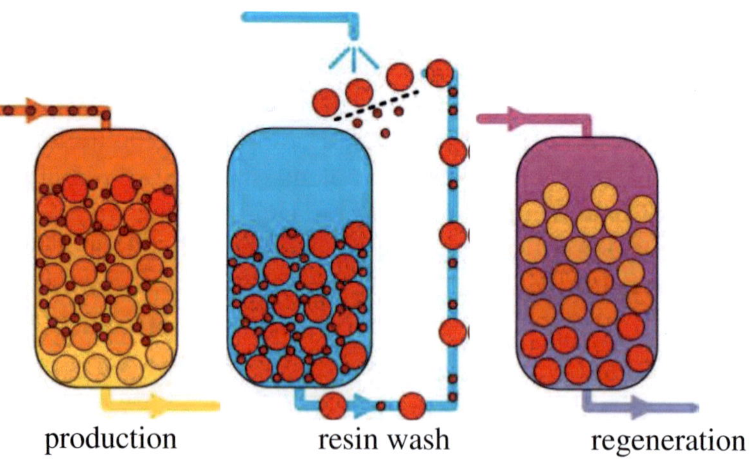

Figure 6.1 System for treating cloudy juices (Courtesy of Bucher Unipektin)

Bucher Foodtech developed a system for unfiltered juices containing up to 1% pulp, where the adsorbent after the loading cycle is transferred out of the column, washed to remove the retained particles and then put back into the column for regeneration (fig. 6.1).
Figure 6.2 shows a 2-vessel citrus debittering system equipped with resin washing drum for cloudy juices.

Figure 6.2 Two-vessel citrus debittering plant for cloudy juice applications (Courtesy of Bucher Unipektin)

The adsorbent used in orange juice debittering, or in removing naringin from grapefruit juice, is of styrene-DVB type. The suggested specific flow rate is 5 BV/h. The adsorbent resin removes limonin to a large extent and toalesser extent flavonoids like hesperidine and polyphenolic compounds. At this specific flow rate, cycle time is between 6 and 12 hours, longer if only limonin is removed. Regeneration is achieved with NaOH, as described in an earlier section (p.66-67) with the following differences (Milnes and Agmon, 1995). Regeneration temperature recommended is >93°C in order to sterilize the adsorbent column. An additional step is added after the NaOH injection and the water rinse that follows: a dilute solution (0.1%) of H_2O_2 is allowed to pass through the adsorbent. This chemical oxidizes irreversibly adsorbed molecules and allows their easier removal from the adsorbent. After H_2O_2, it follows a water rinse step and then the acid injection which here is H_3PO_4 which is already found in orange juice so that the acid wash does not alter the juice composition. Care must be taken in the H_2O_2 concentration not to exceed 0.1% otherwise H_2O_2 may oxidize the adsorbent thus shortening its life time.

If it is also desired to have in addition to debittering a deacidification unit, this is positioned downstream from the adsorbent. As discussed above, because at the begining of the loading cycle the pH of the deacidified juice can go higher than 4.6, the effluent juice from the WBA resin is blended with non-deacidified juice to balance the pH. Therefore only part of the debittered juice is deacidified, the rest being used to balance the pH. After deacidification and blending with the non-deacidified

188

juice, the pulp is put back to the clarified juice in the desired proportion (figure 6.3).

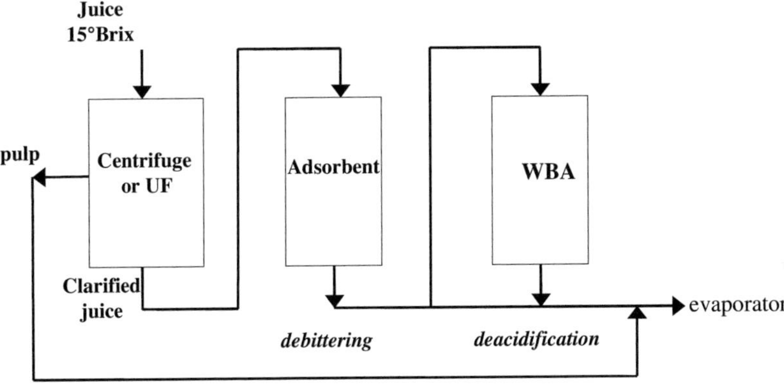

Figure 6.3 Debittering and deacidification of orange juice with IER

Other fruit juices

Grape juice contains about 10-15% carbohydrates. Purification of this juice, as other fruit juices, provides a substitute of sugar that can be added into fruit drinks and carbonated beverages. Grape juice that has been decolorized,demineralised and concentrated can be added into grape must before fermentation if the sugar content of the must is not high enough to make high quality wine.

189

Grape juice purification can be achieved with cation and anion exchange resins as discussed already in apple juice demineralization in page 148. The grape juice contains typically more anions than cations and therefore there is more anion exchange resin than cation exchanger. The main organic acids in a grape juice ar tartaric and malic. Tartaric acid is found as free acid and as potassium bitartrate. Except potassium which is the main mineral in grape juices, other ions include sodium, iron, phosphates, sulphates and chlorides. An IER layout is as follows:

$$SAC \rightarrow WBA \rightarrow SBA \rightarrow SAC$$

The WBA and SBA can be in two or in one vessel as a stratabed configuration[12]. The first SAC removes cations such as K^+, Ca^{2+} and Mg^{2+}, the WBA removes the strong acids while the SBA removes the weak acids and the last SAC removes any nitrogenous materials and also any alkalinity that leaked from the previous resins. In some flow sheets, there is an adsorbent before the first SAC resin. The exact resin volumes depend of course on the grape juice composition. Typical figures are that the WBA has about double the volume of the SBA resin and the ratio of the WBA+SBA resins to the first SAC is about 1.7. The last SAC has about half the volume of the first SAC. The specific flow rate is about 2 BV/h where one BV is the volume of WBA+SBA resins. The cycle length for a grape juice having 15°Bx and about 180 meq/L anions is about 3 hours.

---

[12] In a stratabed, the WBA resin is on top of the SBA resin.The particle size of the two resins is such that by fluidization the two resins separate due to their density difference, the WBA being lighter goes up and the SBA remains in the lower part of the column. Upon settling, the two resins are clearly separated.

Regeneration is performed using 100 g H_2SO_4/L_R for co-current regeneration of the cation exchangers, going from the last to the first SAC and 90 g $NaOH/L_R$ for the anion exchangers going from the SBA to the WBA resin. The main anion fixed on the anion excha,gers is tartrate. The solubility of di-sodium tartrate (Na2T) at ambient temperature is 0.1M. Therefore, the NaOH concentration of the regenerant should be appropriate to avoid Na2T precipitation during anionic regeneration.

Another system promoted by Eurotec WTT[13] is the following:

$$WBA \rightarrow SAC \rightarrow WBA/SBA \rightarrow SAC$$

The first WBA fixes the free acidity, the SAC removes cations, the second WBA with the SBA remove the acidity associated with the cations and the final SAC resin acts as a polisher.

Special regeneration sequences are proposed in this technology: H_2SO_4 passes from the first WBA to the WBA/SBA column and collected to recover weak acids. The first SAC is regenerated with H_2SO_4 or HCl and the regenerant goes from the SAC to the first WBA resin. This operation cleans the WBA resin from absorbed organics and converts it to the Cl^- or SO_4^{2-} form. The spent regenerant goes to waste. The anion resins are regenerated with 4% NaOH going from SBA/WBA to the first WBA and to the wastes.

Pear juice is another fruit juice that after decolorization and demineralization can be used as a blending juice for making fruit drinks. The purification of pear juices is achieved in a similar way as the apple juice discussed above. The layout is:

---

[13] WTT Water Treatment Technologies

$$SAC \rightarrow WBA \rightarrow adsorbent$$

Pinapple mill juice is the juice obtained by pressing the shells and other waste parts of pinapples. This juice is heated to precipitate albumins, limed to precipitate citrates and filtered. The filtered juice, which is essentially a sugar solution, is then demineralised to be used as canning syrup (Mindler, 1949; Ashurst 2002).

The demineralization is done with three pairs of SAC$\rightarrow$WBA resins in a merry-go-round configuration, two pairs on loading in series and one on regeneration. The cycle stops when the first pair is exhausted. Then the exhausted pair goes to regeneration, the second pair becomes the head columns and a freshly regenerated pair goes to tailing position.

$$SAC \rightarrow WBA \rightarrow SAC \rightarrow WBA$$

The exhausted resins, after sweetening off, are treated in series, from the SAC to the WBA resin, with 10% NaCl solution. The Na^+ ions are exchanged for the Ca^{2+} fixed by the SAC. The brine solution containing now Ca^{2+} passes through the anion exchange resin and removes citrate ions fixed by the WBA. The calcium citrate is recovered from the spent brine solution by heating after which the calcium citrate precipitates out and the NaCl is recycled. After the NaCl treatment, the regeneration of the resins is done with H_2SO_4 for the SAC and with NaOH for the WBA resin.

Beverages

Beer

Haze formation in beer has a similar mechanism as that for apple juice (page 182). It is due to the interaction between proteins containing proline, and polyphenols (Siebert, 1999). Beer contains proportionally more haze active proteins than HA polyphenols contrary to the apple juices that contain more HA polyphenols than HA proteins. Haze active proteins can be removed by silica which binds HA proteins in a similar manner as HA polyphenols do.

Synthetic adsorbents and ion exchange resins have been tried in beer and wine stabilization from haze formation (Bellamy and Pease, 2010). One of the objectives of these trials was to remove proteins to stabilize from haze formation without removing too much color from the treated wine or beer. An adsorbent functionalized with sulfonic groups, Amberlite® FPX62, in H^+ or in K^+ form showed good stabilization results with relatively low decolorization of beer or wine. Inert adsorbent Amberlite® FPX 66 showed too high color removal and therefore was not suitable for this application.

Decaffeinated coffe or tea.

Some plants such as coffee or tea, contain caffeine which gives certain properties to the drinks made from these plants. The main properties of interest are stimulation of the nervous system and a diuretic action. However, these properties are not

always considered as beneficial and in some cases are
undesirable, hence the inerest to produce decaffeinated coffee or
tea.

Structure of caffeine

Decaffeination of coffee in most cases takes place by extraction
of the green coffee beans. Organic solvents used in this process
include methylene dichloride or trichloroethylene. The process
consists in increasing the moisture of the green beans, extract
the desired amount of caffeine, steaming the extracted beans to
remove the organic solvent and removing the excess moisture.
Caffeine is recovered by distilling off the solvent and then it
can be used either in medical uses or as an additive for other
drinks that they do not contain caffeine.

In order to avoid using organic solvents, an early patent by
Nestlé SA (Lausanne, CH) suggested the removal of caffeine
from coffee or tea extracts with a synthetic polymeric adsorbent
of styrene-DVB of Amberlite® XAD4 type (Margolis *et al*,
1977).

Later, a popular process was developed where the extraction of
caffeine from green coffee beans is done with supercritical CO_2
as a solvent. This process has the advantage of being moved

194

away from organic solvents and their perceived dangers in their use in food products. The green beans are steamed and then added to a high pressure vessel. A mixture of liquid CO_2 and water is forced through the coffee beans at a pressure of 300 atm and a temperature of 150°F. Caffeine dissolves in CO_2 while compounds contributing to the odor and taste are insoluble and remain in the beans. Caffeine is recovered from the caffeine-rich CO_2 by washing with water followed by a RO to recover caffeine. An option to that is to use a strong acid cation exchanger to adsorb out the caffeine from the CO_2 stream (Clarke RJ and Vitzthum OG, 2008). The resin can be regenerated with salt or with acid from where caffeine is recovered by concentration and crystalization or by solvent extraction.

7, Drinking water purification

The quality of drinking water in the European Community, including process water for food processing, is ensured by the Drinking Water Directive which defines all the parameters that must be monitored and tested regularly. World Health Organization's guidelines are used as a basis.
In many of these parameters, IX rechnology is a suitable means for making drinking water safe for human consumption. These technologies are described in this section..

Nitrates removal

The nitrate nevel in natural waters varies usually between 5 and 15 mg NO_3/L. It happens frequently nevertheless that nitrates are found at higher concentrations due mainly to fertilizers. In the year 2001, the upper limit for nitrates in drinking water was set at 50 mg/L; For infants and pregnant women, the limit is much lower, as an indication it is 5 mg/L.
Nitrates can be removed with conventional strong base anion exchange resins Type 1 or Type 2. These resins however, have higher affinity for sulfate ions than for nitrate. Consequently, if the sulfate concentration in the water ts high, the operating capacity of the resin for nitrate removal is decreased. In addition, because of chromatographic peaking (figure 2.11, page

69), if the service cycle is overrun, there is a risk of having nitrates in the treated water with a concentration higher than that of the feed water.

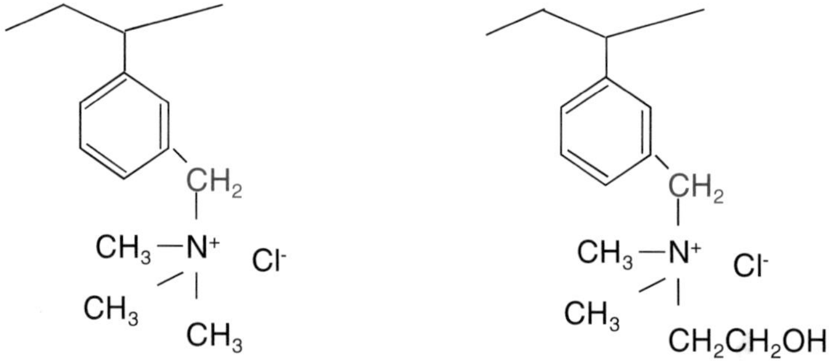

Figure 7.1 Conventional SBA exchange resins

In view of this, resins were developed that have higher selectivity for nitrates than for sulfates. This was achieved by using higher alkyl groups on the N atom of the functional group. These alkyl groups, for steric reasons, make more difficult for a divalent or polyvalent anions to be fixed on the functional group while monovalent anions can be fixed easily.

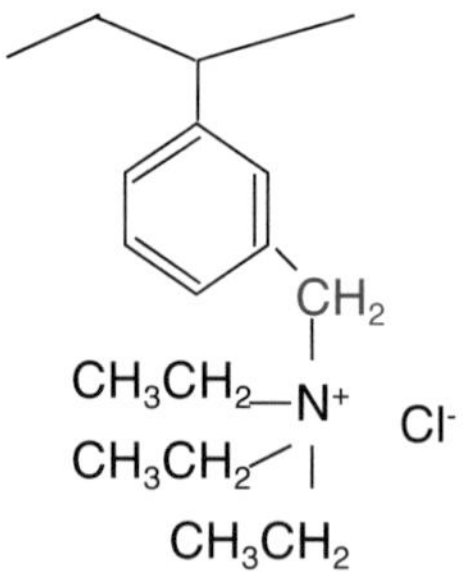

Figure 7.2 Nitrate-selective resins

198

The longer the alkyl groups, the more selective is the resin for monovalent anions over divalent ones. On the other hand, the longer the alkyl groups are, the more difficult is the regeneration of the resin and also the slower are the kinetics of ion exchange in a way that the service cycle becomes too short due to early leakage of the monovalent anion.

For nitrates removal, SBA resins having triethylammonium functional groups and perhaps tripropylammonium groups are in general used, like for example Amberlite™ PWA15, Lewatit® SR7 or Purolite® A520E. Regeneration of the resins is done with NaCl solutions, usually in counter flow direction.

While the use of ion exchange in nitrates removal is a simple process, the spent regenerant disposal can become a problem. Spent regenerants contain high nitrate concentration and still significant quantities of NaCl. Some technologies have been developed to face this problem, one of which being the circulation of the spent regenerant solutions through a biological denitrification reactor where bacteria in the reactor convert nitrates to nitrogen gas.

The CARIX process

The CARIX (**CA**rbon dioxide **R**egenerated **I**on e**X**changers) process consists of a mixed bed of a WAC exchange resin of methacrylic acid type in the H form and a SBA exchange resin in the HCO_3^- form. It is used in municipal water treatment for reduction of calcium, sulfates and nitrate concentrations.

During service, the WAC resin reacts with salts like $CaCl_2$:

$$2\,R\text{-}COO^-H^+ + CaCl_2 \leftrightharpoons (R\text{-}COO^-)_2Ca^{2+} + 2\,HCl$$

The presence of the SBA resin in the HCO_3^- form at an intimate distance, removes rapidly the formed HCl:

$$R\text{-}N^+HCO_3^- + HCl \rightarrow R\text{-}N^+Cl^- + H_2CO_3$$

The overall ion exchange reaction is therefore:

$$\begin{array}{l} 2\,R\text{-}COO^-H^+ \\ \quad\quad\quad + CaCl_2 \rightarrow \\ 2\,R\text{-}N^+HCO_3^- \end{array} \quad \begin{array}{l} (R\text{-}COO^-)_2Ca^{2+} \\ \quad\quad\quad + H_2CO_3 \\ 2\,R\text{-}N^+Cl^- \end{array}$$

Regeneration is done simultaneously for the WAC and the SBA by introducing a carbon dioxide solution under pressure.

$$\begin{array}{l} (R\text{-}COO^-)_2Ca^{2+} \\ \quad\quad\quad + H_2CO_3 \rightarrow \\ 2\,R\text{-}N^+Cl^- \end{array} \quad \begin{array}{l} 2\,R\text{-}COO^-H^+ \\ \quad\quad\quad + CaCl_2 \\ 2\,R\text{-}N^+HCO_3^- \end{array}$$

As seen from the loading and regeneration equations, the wastes in the spent regenerant are the same as the salts removed during loading, that is there are no additional salts in the wastes, making the CARIX process friendly for the environment.

Perchlorates and pertechnetates removal

Perchlorates, ClO_4^-, is a constituent of rocket fuels and has been found in many underground waters especially in the USA. The maximum limit in drinking water has been set at 25 ppb but many States have set lower limits.
Pertechnetates, TcO_4^-, originates from radioactive operations. The limit for drinking water is 1 ppb.
For both these elements, strong base anion exchange resins can be used but, as with the case of nitrates removal, the very low fraction in the feed water and the very low limits allowed in the treated water makes conventional SBA resins unsuitable. By increasing the length of the alkyl cains of the functional groups increases the selectivity for ClO_4^- and TcO_4^- but at the same time, regeneration with NaCl becomes inefficient and also the kinetics of ion exchange become too slow.

In 2001, Oak Ridge National Laboratory (ORNL) developed a regeneration system using $FeCl_4^-$, formed from FeCl3 and HCl. The selective resins for ClO_4^- prefer $FeCl_4^-$ from ClO_4^- and rherefore $FeCl_4^-$ displaces ClO_4^- from the resin (Gu et al, 2001). After regeneration, the $FeCl_4^-$ is displaced from the resin wirh water whereby the concentration of Cl^- drops and the complex $FeCl_4^-$ is converted to some nutral or positively charged Fe compound that desorbs from the resin. After water rinse, the resin is found in the Cl^- form ready for the next cycle.

In order to obtain a compromise between long alkyl chains and kinetics, a new type of SBA resins was developed at ORNL (Alexandratos et al, 2000; Gu and Brown, 2006). These resins have two kinds of functional groups, one with triethylammonium and another with trihexylammonium groups. The presence of trihexylammonium ensures high selectivity and the presence of triethylammonium ensures fast kinetics. These bifunctional resins regenerate very efficiently with $FeCl_4^-$ although they can also be used as non-regenerable resins due to very long service cycles.

Chromates removal

The most frequently found forms of chromium are chromium (III) and chromium (VI) (chromates). Cr(III) occurs naturally while Cr(VI) is essentially produced by certain chemical processes. Cr(VI) is particularly toxic. In drinking waters, the maximum allowed levels are 50 ppb for total chromium with tendency to go lower, to10 ppb.

Cr^{3+} is removed from waters with a SAC resin. Cr(VI) is encountered as chromates, CrO_4^{2-} and as dichromates, $Cr_2O_7^{2-}$. In aqueous solutions, chromates and dicromates exist in equilibrium:

$$CrO_4^{2-} + H+ \rightleftharpoons HCrO_4^-$$
$$2\ HCrO_4^- \rightleftharpoons Cr_2O_7^{2-} + H_2O$$

From these two equilibria, a third one can be derived:

$$2\ CrO_4^{2-} + 2\ H^+ \rightleftharpoons Cr_2O_7^{2-} + H_2O$$

In acidic conditions, the dichromate can be the predominant species, depending on the total Cr(VI) concentration, while in alkaline conditions the chromate in the predominant species. The chromates and dichromates are rather strong oxidizing agents. In acidic pH, Cr(VI) is reduced to Cr^{3+}:

$$Cr_2O_7^{2-} + 14\ H^+ + 6e^- \rightarrow 2\ Cr^{3+} + 7\ H_2O \qquad E°=+1.33\ V$$

In alkaline pH it gives chromium hydroxide:

$$CrO_4^{2-} + 4\ H_2O + 3\ e^- \rightarrow Cr(OH)_3 + 5\ OH^- \qquad E°= -0.12\ V$$

The more positive the redox potential $E°$ is, the greater the affinity of the species for electrons and the tendency to be reduced is. Consequently, in acidic solutions, chromates are stronger oxidizing agents than in alkaline solutions.

Removal of chromates has been achieved with strong (Oberbofer, 1965) or weak base (Oberbofer, 1968) anion exchangers. For Cr(VI) removal from cooling waters blowdown or from ground waters, weak base anion exchangers are used in acidic pH. For removing low levels of chromates from drinking waters at neutral pH, strong base anion exchangers are used.

The removal of chromates by anion exchange resins has been studied in depth (Sengupta and Clifford, 1986). Operating capacities at low pH were found to be much higher than at alkaline pH and therefore, acidic pH is generally used for chromates removal by anion exchange resins. The selectivity of strong base anion exchangers for chromates versus chlorides or sulfates was found to be very high, especially with styrene-DVB resins compared to acrylic resins. This is due to the hydrophobic nature of chromates and of the Sty-DVB resins. In addition, the presence of both, CrO_4^{2-} and $Cr_2O_7^{2-}$ on the anion exchanger increases the selectivity of these resins for chromates.

Regeneration of WBA exchangers is done with NaOH and this allows the recycling of Na_2CrO_4 from the spent regenerant. SBA exchangers need NaCl as regenerant with the addition of some NaOH to convert dichromates to chromates which results in an easier elution. Because at high ionic strength, the selectivity of the resin for Cl^- over CrO_4^{2-} is reversed, the elution of CrO_4^{2-} by Cl^- is achieved easily.

The isotherms chromate-sulfate and chromate-chloride at acidic pH (4.0) showed for both weak and strong base anion exchangers an unfavorable equilibrium at low chromate concentrations and favorable equilibrium at high concentrations, with an inflection point in-between. This inflection point occurs at a chromate concentration of about 4-5 mg/L. The consequence of this is that the breakthrough curves show an early gradual breakthrough of chromates.

Weak base anion exchangers need to be in the protonated form, with HCl or H_2SO_4, to be able to fix chromates. The feed solution pH is adjusted to a pH of about 4. If the pH is lower, below 3 for example, the Cr(VI) may be reduced to Cr^{3+} with resin oxidation and formation of chromic oxide ($Cr(OH)_3$) which precipitates on the resin.

Removal of chromates from drinking water has also been achieved with an acrylic SBA exchange resin bearing quaternized diethylenetriamine functional groups (Reckziegel *et al,* 2018). The removal of chromates takes place at close to neutral pH and the resin showed considerably higher chromate removing capacity compared to a commercial SBA resin used in chromate removal.

As mentioned, WBA exchangers can remove chromates from the water under acidic conditions. One WBA resin that showed a very high loading capacity for chromates is the formophenolic WBA Duolite® A7 (or Amberlite^{TM} PWA7 of Dow Chemicals, now DuPont) (Sarkal et al, 2016). The feed water had 120 µg/L Cr(VI), 50 mg/L Cl, 50 mg/L SO_4^{2-}, 10 mg/L HCO_3^- and a pH of 5 while the effluent pH started at about 6.5 indicating that during loading, the resin consumed acidity. Breakthrough took place after 12000 BV. The extremely high chromate removing capacity of Amberlite PWA7 was attributed ro a redox reaction where Cr(VI) oxidizes the matrix of this resin, rather than the functional groups, where Cr(VI) is reduced to Cr(III) and precipitates in the resin, thus liberating exchange sites for more chromates removal.

The removal of traces of chromates from water containing other salts at higher concentrations is conceivable because the resin is much more selective for chromates compared to other anions like Cl^- or SO_4^{2-} (see pages 36-37). For example, with a water containing 10-20 ppb CrO_4^{2-}, and some 50 mg/L of Cl^-, SO_4^{2-} or NO_3^-, the cycle length for CrO_4^{2-} removal at neutral pH with non-acidified feed solution, can be very long.
In fact, such cases have been reported where the resin Amberlite^{TM} PWA7 at about neutral pH has given extremely long service cycles in removing CrO_4^{2-} (Romano, 2018). The water composition was pH=7.7 Cl=12 mg/L, NO_3=48 mg/L, SO_4=34 mg/L Cr(VI)=15.4 µg/L.

.

A different approach for chromates removal from contaminated waters is with a selective resin bearing a chelating functional group where a metal, such as Cu(II), is fixed (SenGupta and Zhao, 1998). This is an example of ligand exchange where Cu(II) removes selectively chromates by complex formation. With a feed solution containing 0.1 mg/L Cr(VI), 2 meq/L SO_4^{2-}, 1 meq/L HCO_3^-, 1.5 meq/L NO_3^- and 1.5 meq/L Cl^- at a pH of 7.5 and at 20 BV/h, the Dowex® M4195 in Cu(II) form treated about 7500 BV to breakthrough compared to about 300-400 BV treated by a strong base anion exchanger Amberlite® IRA900 Cl.

Arsenic removal

Arsenic is found in water as an anion, either as trivalent, forming arsenous acid As(OH)$_3$ and arsenites, or as pentavalent as arsenic acid, H_3AsO_4 and the arsenate salts. Arsenous acid has pK_{a1}, pK_{a2} and pK_{a3} values of 9.23, 12.13 and 13.40 respectively and therefore in water it is found as non-dissociated at pH below about 9 and for that reason As(III) is more difficult to remove than As(V). Arsenic acid has pK_{a1}, pK_{a2} and pK_{a3} values of 2.2, 7.0 and 11.5 respectively and therefore in water at pH about 6-8 it is found as as $H_2AsO_4^-$ and $HAsO_4^{2-}$.
Arsenic is most frequently removed by first oxidation of As(III) to As(V) followed by coagulation-flocculation process with alum, ferric chloride or ferric sulphate. Other technique include the co-precipitation of As(V) with Fe(III)-oxides where the soluble As(V) is incorporated in the growing Fe(III) hydroxide

206

phase, and adsorption on activated alumina (Al_2O_3), hydrated ferric oxide, activated carbon and other adsorbents.

Adsorbents based on granular ferric hydroxide are also used to remove arsenic, vanadium, selenium, antimony, molybdenum, along with other metals such as chromium, uranium, copper and lead.

Arsenic (III) reacts with thiol groups to form As-S bonds. It is possible therefore to use thiol resins or other supports bearing thiol groups to fix As(III) (Tongesayi, 2014).

Ion exchange resins impregnated with ferric hydroxide have been synthesized for the removal of oxoanions (arsenates, chromates, phosphates, oxalates, phthalates). Although strong acid cation and strong base anions have been used as supports for ferric hydroxide, anion exchange resins showed higher capacity due to the fact that the oxoanions are not excluded by the Donnan phenomenon by the strong base anion exchangers (SenGupta and Cumbal, 2007). Similarly, SBA exchange resins impregnated with hydrated zirconium oxide were synthesized for similar applications (SenGupta and Padungthon, 2013). Lewatit® FO36 is an example of a WBA resin filled with iron oxide particles. Both As(III) and As(V) are removed. Regeneration is done with NaOH.+ NaCl solution. FerrIX^{TM} A33E of Purolite is another example of this kind of hybrid anion exchange resins bearing iron oxide nanoparticles for As removal.

Arsenic can be removed by conventional SBA resins (Berdal et al, 2000). Since arsenous acid is non-dissociated at pH<9, As(III) must first oxidized to As(V) as arsenic acid, H_3AsO_4 in order to be removed by IX at about neutral pH. The senectivity sequence of conventional SBA resins are as follows:

$$SO_4^{2-} > HAsO_4^{-} > NO_3^{-} > Cl^{-}$$

Therefore, the feasibility of IX process depends on the water composition and in particular the SO_4^{2-} concentration. If SO_4^{2-} are found in concentration >50 ppm, IX may not be an economical solution. Regeneration is done with NaCl brine. Generally, conventional IX resins have lower capacities compared to adsorption on iron-impregnated resins.

One of the primary concerns related to IX treatment is the phenomenon known as chromatographic peaking, (figure 2.11, page 69), during which As(V) levels in the effluent exceed those in the influent stream. This can occur if sulfate are present in the raw water and the bed is operated past the As(V) breakthrough. The SO_4^{2-} being stronger fixed by the resin, displace As resulting in an effluent concentration higher than the feed.

In order to avoid this peaking, the service cycle should be stopped at a given As(V) value and before the breakthrough of sulfates.
 Arseniates and other oxoanions can form complexes with some metals among which Fe^{3+} (see page 35). Thus, chelating resins in the Fe^{3+} form can remove arseniates from water by ligand exchange (Chanda *et al*, 1988 and 1988a). Similarly, phosphonic resins (Diaion® CRP200) in Zr(II) form removes arsenic from water by a similar mechanism (Jyo et al, 2005).

Phosphates removal

Iron (hydr)oxides have the ability to adsorb phosphates, arsenates and uther oxoanions from water due to the formation of complexes with iron by ligang exchange and also due to electrostatic attraction.
Although conventional SBA exchangers can remove phosphates, they are not selective enough to remove down to very low levels in the presence of other anions, especially sulphates for which SBA resins have higher affinity than phosphates.

Selective resins for phosphates removal from drinking water have been developed by binding ferric uxides nanoparticles on anion exchange resins (Blaney *et al*, 2007; Martin *et al,* 2009). Desorption of phosphates from the resin can be achieved with alkaline solutions.

 Phosphates can also be removed from water using ligand exchange resins. These are resins bearing ligand groups grafted on the resin matrix and where a transition metal is fixed on. Phosphates, as other oxoanions, interact with the metal of the resin by Lewis acid-base interactions forming coordination bonds. An example of a commercially available resin is the DowexTM M4195 in Cu2+ form. Since the ligand of the functional group is bispicolylamine that has no electrical charge, the phosphates are fixed on the resin not only with Lewis acd-base interactions but also with electrostatic forces so that this resin has a very high affinity for phosphates.

Fluorides removal

Fluorides can be removed from waters by adsorption on activated alumina, by aluminum coagulation or by reverse osmosis. An ion exchange technology has been introduced (Popat *et al*, 1994; Oke et al, 2011) using an AMP chelating resin in the Al^{3+} form.

This resin gave an operating capacity of about 3-6 g F/L_R depending on the pH of the feed solution. Slightly acidic solutions gave the highest capacity. Specific flow rate was 10 BV/h with a feed solution containing 127 mg F/L without any other salt background. It is interesting to know that the operating capacity of the resin increases as the salt background of the feed solution inceases. This was explained as due to changes in the ligands around the Al^{3+} atom at different salt backgrounds. Regeneration of the resin was done with 3 BV of a 5.5% $AlCl_3$ solution.

Boron removal

Drinking water may contain up to few mg/L of boron. The recommended by WHO upper limits are 0.3 mg/L. Removal of boron from water can be done with boron-selective resins, which are WBA exchange resins bearing a methyl glucamine group:

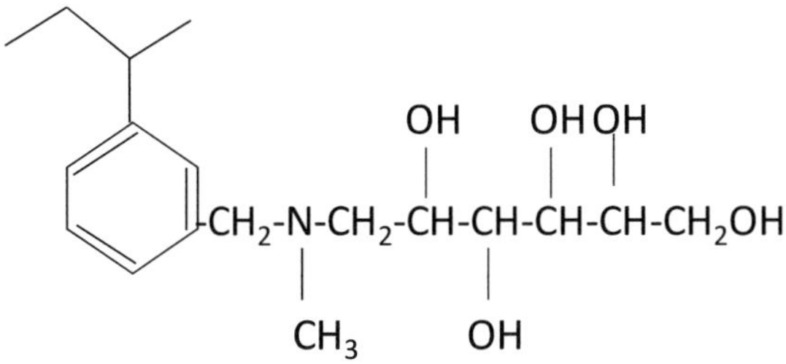

Boron in water is found as boric acid, H_3BO_3, which is a very weak acid and with water it forms borate anions:

$$H_3BO_3 + H_2O \rightarrow [B(OH)_4]^- + H^+$$

Boric acid reacts with polyols to form anionic complexes:

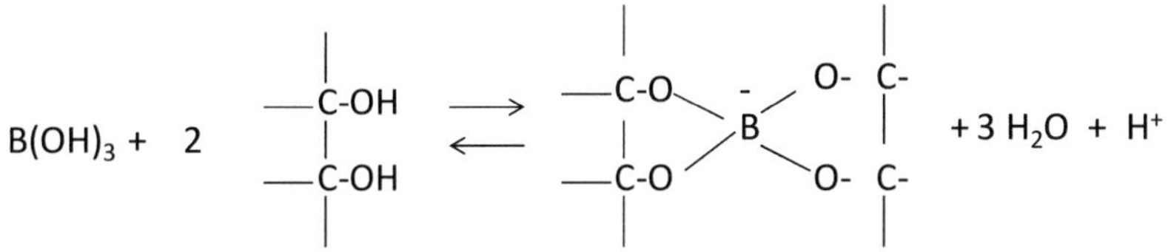

The stability of these complexes depends on the pH. At low pH the above equilibrium shift to the left and the complex dissociates.

The removal of boron with boron-selective resins is based on the above chemistry. Loading is done at a neutral pH. Regeneration is done with H_2SO_4 or HCl acid. In order to buffer the pH back to neutral for the following service cycle, a caustic wash is done after the acid regeneration.

An illustration of the expected operating capacity for boron removal from drinking water is given in figure 7.3:

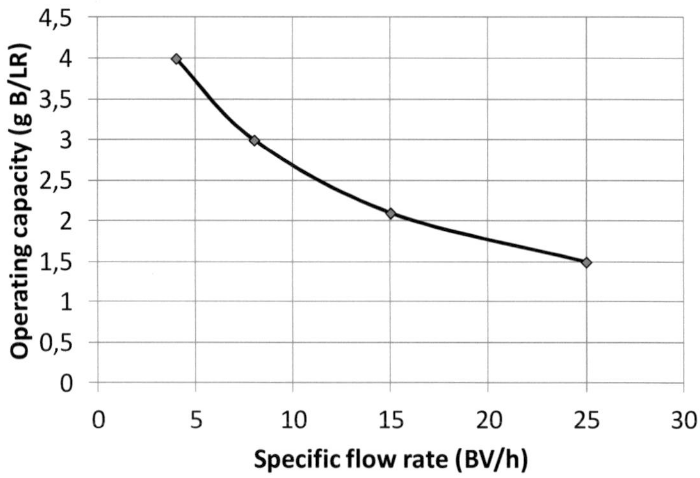

Figure 7.3   Boron removal operating capacity as a function of
specific flow rate

The boron-selective resin is essentially a styrene-DVB WBA exchanger which however, as many styrenic WBA resins, have a certain quaternary ammonium functional groups. After the caustic wash of the regeneration step, the WB groups are in the free base form while the SB groups are in the OH^- form and some remaining SO_4^{2-} form. During the loading period therefore, except the removal of boron, there may be some exchange of the OH^- or SO_4^{2-} ions for an equivalent amount of ions from the feed water. Consequently, there may be some composition changes in the treated water.

212

Uranium removal

In drinking water in the presence of CO_3^{2-} ions, uranium is found as carbonate complexes, as indicated in the following table:

pH	3	5	7	9
Species	UO_2^{2+}	UO_2CO_3	$UO_2(CO_3)_2^{2-}$	$UO_2(CO_3)_3^{4-}$

Removal of uranium from drinking water is done wirh strong or weak base anion exchange resins.
In drinking water, uranium levels vary. Typically they do not exceed the 10 µg/L but in some cases they may be as high as 100 µg/L.

Strong base anion exchange resins are suitable to remove uranium from drinking water. In a study (Hanson *et al*, 1986) it was demonstrated that different commercial SBA resins were able to remove uranium from water from a feed concentration of 200-400 µg/L down to <14 µg/L for 12000 to 14000 BV. Regeneration was done with NaCl.

Acrylic and styrenic weak base resins have also used for uranium removal from ground water (Riegel and Höll, 2008). Acrylic gave a beter loading capacity, possibly because acrylic WBA resins are a little stronger bases than styrenic. One important factor, as expected, was the pH of the feed solution. At pH=7-7.5 and a uranium concentration of 1000 µg/L in carbonated water, the acrylic resin (Amberlite[TM] IRA67 of Rohm and Haas Company) treated more than 30000 BV before breakthrough.

8. Aminoacids separation and purification

Aminoacids have the general formula:

$$H_2N-CH-COOH$$
$$|$$
$$R$$

They have a carboxylic acid group and an amine group attached to a tetrahydral methylene carbon. The different aminoacids are distinguished by the side chain R attached on the methylene carbon.

At low pH, the carboxylic acid and the amine groups are both protonated. The aminoacid has then a net positive charge. As pH increases, the carboxylic group dissociates yielding a zwitterion with zero net charge (8.1). At even higher pH the amine group deprotonates to give the negatively charged carboxylate ion (8.2):

$$H_3N^+CH(R)COOH \rightleftharpoons H_3N^+CH(R)COO^- + H^+ \qquad K_{a1} \quad (8.1)$$

$$H_3N^+CH(R)COO^- \rightleftharpoons H_2N-CH(R)COO^- + H^+ \quad K_{a2} \quad (8.2)$$

where K_{a1} is the dissociation constant for the carboxylic acid group and K_{a2} the dissociation constant for the amine group.

For an aminoacid containing only one carboxylic acid group and one amine group, at pH values above the pK_{a1} the negatively charged carboxylate group predominates. At pH values below the pK_{a2}, the positively charged ammonium form predominates. At a pH below the pK_{a1}, the predominant form will be an undissociated carboxylic acid and a positively charged ammonium group (eq.8.1). At a pH above the pK_{a2}, the predominant form will be an undissociated amine group and a negatively charged carboxylate group (eq.8.2). At pH values between the two pKa's, the predominate form will be a negatively charged carboxylate group and a positively charged ammonium group, called zwitterion. At exactly the midpoint between the pKa values, the positive and negative charges balance out completely and the aminoacid has zero net charge. This point is called isoelectric point. In the rest of the pH values between the two pKa's, the aminoacid will have a small net positive or net negative charge. The pH at the isoelectric point, pI, can be calculated from the relationship: $pI=1/2(pK_{a1}+pK_{a2})$.

Some aminoacids contain additional carboxylic or amine groups in the side chain R. If the side chain contains a carboxylic acid group with a pK_{aR}, then the pI is calculated from: $pI=1/2(pK_{a1}+pK_{aR})$. If the side chain contains an amine group with a pK_{aR}, then the pI is calculated from: $pI=1/2(pK_{aR}+pK_{a2})$.

For example, lysine (Lys, MW=146, figure 8.1) has an amine group in the side chain with a pK_{aR} of 10.67. The other two ionizable groups have $pK_{a1}=2.15$ and $pK_{a2}=9.16$. The pI is ½(9.16+10.67) = 9.92. At a pH about one, Lys is found with the two amine groups in the protonated form (fig. 8.1 (A)). At a pH

216

close to the pI, the carboxylic acid is in the negatively charged carboxylate form and the side chain amine groups in the positively charged ammonium form so that the net charge is zero (fig. 8.1 (C)). At a pH about 5, Lys has both amine groups positively charged and the carboxylic acid group negatively charged (fig. 8.1 (B)). Above the pK_{a2}, at about pH=11, the two amine groups are non-ionized while the carboxylic acid group is in the carboxylate form (fig. 8.1 (D)).

$$H_3N^+\text{-CH-COOH} \qquad H_3N^+\text{-CH-COO}^- \qquad H_2N\text{-CH-COO}^- \qquad H_2N\text{-CH-COO}^-$$

$$(CH_2)_4 \qquad\qquad (CH_2)_4 \qquad\qquad (CH_2)_4 \qquad\qquad (CH_2)_4$$

$$N^+H_3 \qquad\qquad N^+H_3 \qquad\qquad N^+H_3 \qquad\qquad NH_2$$

$$\text{(A)} \qquad\qquad \text{(B)} \qquad\qquad \text{(C)} \qquad\qquad \text{(D)}$$

Figure 8.1 Lysine

Ion exchange resins can be used in aminoacids production from protein hydrolysates or from fermentation broths. All types of ion exchange materials can be used: strong or weak cation exchange resins, strong or weak anion exchange resins or synthetic adsorbents but more frequently are used SAC resins and synthetic adsorbents.

SAC resins have a sulfonic functional group and are therefore charged negatively. They have a styrene-DVB matrix and come in various DVB contents so that the network structure is more or less loose and the diffusion of the various aminoacids and eventually di- or higher peptides is controlled by the DVB content. Depending on the mobile cation, the sulfonic SAC resins come in the H^+ form, Na^+ form, NH_4^+ form or other form. In the H^+ form, the pH inside the resin beads is strongly acidic while in the other forms is close to neutral. Since at acidic pH the aminoacids are found in the positively charged form, all aminoacids can be fixed on a SAC resin in the H^+ form:

$$\mathcal{R}\text{–}SO_3^-H^+ + R\text{-}CH(NH_2)COOH \rightarrow \mathcal{R}\text{–}SO_3^-\,H_3N^+CH(R)COOH$$

If the SAC resin is in the Na^+ form then neutral or basic aminoacids that are positively charged at neutral pH can be fixed by exchange with the Na^+ ions:

$$\mathcal{R}\text{–}SO_3^-Na^+ + H_3N^+CH(R)COOH \leftrightarrows$$
$$\mathcal{R}\text{–}SO_3^-\,H_3N^+CH(R)COOH + Na^+$$

In a similar way, all aminoacids acidified with a strong acid to a low pH, thus being in the cationic form, are fixed on a SAC resin in the Na^+ form. In fact, the loading capacity of the SAC resin for aminoacids depends strongly on the pH of the feeding solution. For example, the loading capacity of a SAC resin for lysine at pH=1 is about twice that at neutral pH.

Methionine, lysine, threonine and tryptophane are aminoacids used in animal feed. Of these, methionine is produced by chemical synthesis while the other three are produced by fermentation. In general, fermentation broths contain 0.1 up to 10% of the aminoacid (usually only one) depending on the fermentation process, microbial cells, proteins, salts and organic materials like sugars and pigments.

The raw materials for lysine production are beet or cane sugar, molasses or starch hydrolysates for the carbon source and ammonia for the nitrogen source. The growth media also contain minerals and growth factors, often obtained from corn steep liquor. The raw material used in the fermentation may affect the ion exchange resins in terms of fouling for example. Part of the produced lysine is made in the hydrochloric acid form as feed material for animals, especially for pigs and poultry.

The recovery and purification of lysine from the fermentation broth is achieved with cation exchange resins as follows. The broth is acidified with HCl or H_2SO_4 to a pH of 1-3 and centrifuged or filtered on UF membranes to remove cells and other suspended matter. At this pH, lysine is found in the positively charged, ammonium form while the carboxylic acid group is very little ionized. The clarified broth is then passed through a SAC in the NH_4^+ form where the positively charged lysine is exchanged with the NH_4^+ of the resin:

$$\mathcal{R}\text{–}SO_3^- NH_4^+ + Lys^+Cl^- \leftrightarrows \mathcal{R}\text{–}SO_3^- Lys^+ + NH_4Cl$$

The cationic impurities in the feed solution are also fixed on the resin along with some peptides and color, depending on the raw materials of the fermentation.

After loading, there is a rinse step to remove the clarified broth from the resin and then lysine is eluted with an alkaline solution, for example NH_4OH. At the pH of the eluent, lysine is found in the negatively charged form: the amine groups are in the free base form while the carboxylic acid group is in the carboxylate form. It will therefore be released by the resin and the eluted lysine will be found in the eluate at high concentration. In fact, the ammonia consumption will be about quantitative:

$$\mathcal{R}\text{–}SO_3^- Lys^+ + 2\,NH_4OH \rightarrow \mathcal{R}\text{–}SO_3^- NH_4^+ + Lys^- NH_4^+ + 2\,H_2O$$

Other impurities such as K^+ will only be partially eluted with NH_4OH because by exchanging K^+ for NH_4^+, the strong base KOH is formed which inhibits the ionization of NH_4OH and the exchange stops:

$$\mathcal{R}\text{–}SO_3^-K^+ + NH_4OH \xrightarrow{\times} \mathcal{R}\text{–}SO_3\text{-}NH_4^+ + K^+OH^-$$

Some of the K^+ can come off the resin by equilibrating with the NH_4^+ of the Lys-NH_4^+. After the NH_4OH elution of lysine, the resin is found partially in the NH_4^+ form and partially in other ionic forms. This should not affect significantly the following cycle where Lys^+ will be exchanged against NH_4^+ or for example K^+. However, since sulfonic SAC resins have higher affinity for K^+ than for NH_4^+ (page 34), if K^+ accumulates on the resin then a broader breakthrough curve will result with some operating capacity loss of the resin.

Another way to purify lysine is to pass the clarified broth through the SAC resin in the H^+ form. After the ammonia elution, the resin is regenerated with H_2SO_4 and thus most of the cations remaining on the resin are removed. In addition, the affinity of the resin for H^+ is lower than for NH_4^+ and therefore a sharper breakthrough curve will result, under similar operating conditions. In the case that the resin works in the H^+ - NH_4^+ cycling process a very good resistance of the resin to breakdown is needed due to the frequent conversions back and forth from NH_4^+ to H^+ form accompanied by expansions –contractions of the resin and due to the heat of neutralization of H_2SO_4 with NH_4OH that locally develops and which may physically damage the resin.

The eluate of the NH_4OH elution is subsequently concentrated by evaporation where NH_4OH is stripped off, then if lysine chlorohydrate is desired, it is acidified with HCl and Lys.HCl is crystallized and dried.

The above described system is a conventional fixed bed ion exchange system. The use of continuous systems presents the advantages of using less resin and less water, less dilution and less water to evaporate. Figure 8.2 illustrates the continuous systemApplexion® Continuous Ion Exchange technology, patented by Novasep for lysine production.

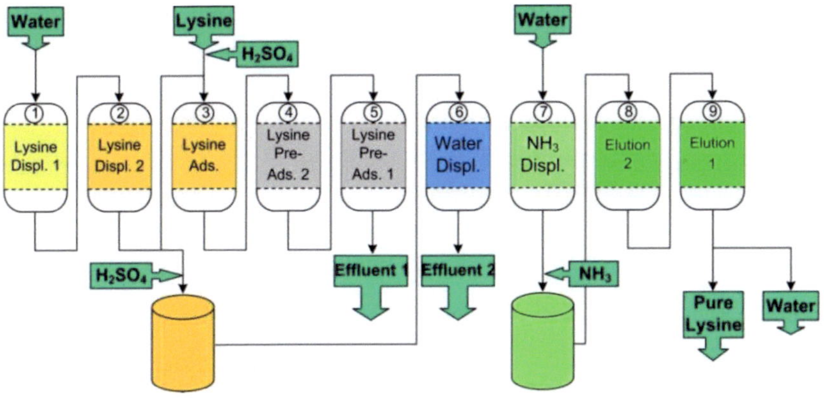

Figure 8.2 Applexion® Continuous Ion Exchange process for the production of lysine (Courtesy of Novasep)

This is a continuous ion exchange process where the inlets and the outlets are shifted in a manner derived from the Applexion® SSMB technology.

Lysine recovery can also be achieved using an SMB system which was discussed earlier (see pages79-80). The resin is a SAC in the NH_4^+ form. The clarified broth enters from the feed port, the ammonia eluent enters from the eluent port and lysine exits from the extract port (Binder, 2001). ISEP$^®$ system was used (Advanced Separation Technologies Inc., a subsidiary of Calgon Carbon Corporation).

A variation of this method was used (Fechter*et al*, 1997; van Walsem and Thompson, 1997) in a ISEP$^®$ system with two sets of columns. The first set contained a SAC resin while the second set a carboxylic cation exchange resin. The clarified and acidified fermentation broth passes through the first set of SAC resin in the NH_4^+ form where Lys^+ is fixed. The elution is done with NH_4OH solution where lysine is eluted in the negatively charged form. This alkaline solution passes then through the second set containing carboxylic cation exchange resins where Lys^- is not fixed being an anion but where cationic impurities are fixed. The effluent is therefore a lysine solution containing only few impurities. This method avoids the crystallization step to produce the final product.

Other continuous systems have been proposed. In one of them (Hitoh*et al*, 1994), the resin comes in contact with the raw fermentation broth batch-wise in contactors under agitation. There is a number of contactors for the loading step, a number of contactors for the elution step and some contactors for washing. The resin moves from one contactor to the next one with a pump while the broth moves counter-current from one contactor to the previous one by overflow. If for example there are four contactors in the loading step, freshly eluted resin comes to contactor #4 and moves towards #3, #2, #1 and then to the elution contactor. Feed solution flows from contactor #1 towards contactor #4 and then to waste. The resin is separated from the broth with screens. In the elution section, if for

222

example there are four elution contactors, exhausted resin goes to contactor 4 and moves through to #1 while the eluent flows counter-current from #1 to #4 and from there to concentration to recover lysine. This system works along the same principles as the Resin in Pulp system used in uranium recovery (Zaganiaris 2017).

 Other aminoacids such as threonine and tryptophane can be recovered in a similar way.
Tryptophane has pKa$_1$=2.38, pKa$_2$=9.39 and pI=5.89. Purification is achieved by passing the fermentation filtrate through a SAC resin where tryptophane is fixed. Elution is done with 0.5N NH$_4$OH and tryptophane is recovered by concentration and crystallization. The SAC exchange resin should have a low degree of crosslinking due to the large molecular size of tryptophane. As was the case with lysine purification, physical stability of the resin is a key issue.

pKa1=2.46   pKa2=9.39   pI=5.89

tryptophane

Glutamic acid is a dicarboxylic aminoacid with the formula

HOOC-CH$_2$-CH$_2$-CH(NH$_2$)COOH

With pK_{a1}=2.16, pK_{a2}=9.58 and pK_{aR}=4.15 the pI is therefore 3.15.

Glutamic acid (Glu) is produced by fermentation. The clarified broth is first passed over a SAC resin in the H^+ form where glutamic acid is fixed as cation. Elution is performed with a NaOH solution where at the high pH of the eluent Glu is found in the negatively charged form and therefore released by the resin as Na^+ salt:

$$\mathcal{R}\text{–}SO_3^-Glu^+ + 2\,NaOH \rightarrow \mathcal{R}\text{–}SO_3^-Na^+ + Glu^-Na^+ + 2\,H_2O$$

The salt is then crystallized from the concentrated eluate. The resin now in the Na^+ form is regenerated with H_2SO_4 to be converted to the H^+ form ready for the following cycle (Das*et al*, 1995).

Monosodium glutamate (MSG) is used as a food additive. Weak base anion exchange resins are used to decolorize MSG. After the service cycle, the resin is regenerated with NaOH solution. In order to minimize MSG losses due to fixation of MSG on the resin, a secondary amine resin, being a weaker base than the tertiary amines, is better. Such a resin is the Duolite® A7 a formophenolic resin from Rohm and Haas Company (now The Dow Chemical Company). However, the long rinse requirements is a drawback. All formophenolic WBA resins require long rinse due to the phenol group that is a weak acid and is converted to the Na^+ form during NaOH regeneration which during rinse is hydrolyzed to give NaOH.

Leucine recovery from fermentation broths (Wu*et al*, 2004) has been achieved after a first crystallization to separate it from other aminoacids produced in the fermentation broth, then redissolving and purifying it on a SAC resin. Elution with ammonia or ammonium chloride.

A number of aminoacids such as serine, threonine, proline, alanine, valine, leucine, isoleucine, phenylalanine, histidine, arginine, are produced industrially from protein hydrolysates. The aminoacids are then separated and recovered using ion exchange chromatography (IEC) with SAC exchange and/or SBA resins, as shown in the diagram below:

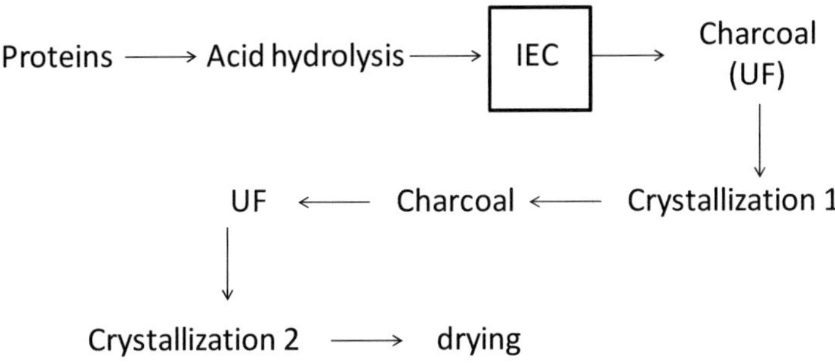

The affinity of the resin for the different aminoacids at a given pH depends upon the charge of the aminoacids at this pH. For example, at about neutral pH, the affinity of a SAC resin for aspartic acid, an acidic aminoacid with a negative net charge at neutral pH, is low, for lysine, a basic aminoacid with a positive net charge at neutral pH is high while for serine, a neutral aminoacid is in-between.

$$H_3N^+\text{-CH-COOH} \qquad\qquad H_3N^+\text{-CH-COO}^- \qquad\qquad H_3N^+\text{-CH-COO}^- \qquad\qquad H_2N\text{-CH-COO}^-$$

CH_2	CH_2	CH_2	CH_2
COOH	COO H	COO$^-$	COO$^-$
(A)	(B)	(C)	(D)
pH=1	pH=2.8	pH=6.6	pH=11
Positive	pI	negative	negative

Fig. 8.3   Aspartic acid

Once these three aminoacids are fixed on a SAC resin, they can be separated by selective elution with a NaCl solution where the Na$^+$ ions displace the aminoacid ions. At the begining, aspartic acid which is only weakly bound, will be eluted first. Then, by increasing the NaCl concentration serine will elute and finally, by increasing further the NaCl concentration, lysine will come off.

Another way to separate aminoacids fixed on a SAC resin is by elutiong with buffers having different pH. As the pH of the buffer eluent changes, so changes the ionic state of the aminoacid and therefore its affinity for the resin. The aminoacids to be separated are loaded on the SAC resin in the chromatographic column at an acidic pH. At these conditions, all aminoacids will be fixed on the resin. Elution is achieved with buffer solutions by steadily increasing the pH. At the pI of the aminoacid there is no net charge and the aminoacid is no longer attracted by the functional group of the resin. Selective elution of the aminoacids is therefore achieved by carefully selecting the pH of the eluent buffer. Temperature, ionic concentration of the eluent and flow rate are also important parameters for the separation.

Lysozyme is a protein (enzyme) having a molecular weigh of 14000-19000 and an isoelectric point of 11.35. Therefore, at pH below 11.3, lysozyme has positive charge. Weak acid cation exchange resins have a pKa values of about 5. This means that at pH below 5, WAC exchangers have no ion exchange capabilities. At pH above 5, these resins are in the carboxylate form and have negative charge.

Hen egg white is a major source for lysozyme. There are different ways to recover lysozyme that have been investigated, one simple method is the use of a macroporous WAC exchange resin like Amberlite^{TM} IRC 50 (Ghlelmetti and Trinchera, 1970) or Duolite^{TM} C-464 (Li-Chan *et al*, 1986) at about neutral pH. The macroporous structure of the resin allows the retatively high molecular weight lysozyme molecules to diffuse into the resin matrix. It should be noted that electrostatic forces are not the only driving forces for fixing lysozyme molecules on the resin. Other interactions probably take place like H-bonding as well as other coordination binding may take place. This is a similar situation in vitamin B_{12} binding on WAC exchange resins mentioned later (page 245).

9. Organic acids

Citric acid

Citric acid is the most important organic acid used in the food industries and finds many applications in foods and soft drinks. It has the formula:

$$HOOC-CH_2-\underset{\underset{\displaystyle COOH}{|}}{\overset{\overset{\displaystyle OH}{|}}{C}}-CH_2-COOH$$

It is a triprotic acid with pKa values of $pKa_1=3.14$, $pKa_2=4.77$ and $pKa_3=6.39$. At a pH below 1.5 citric acid is a non-ionized neutral molecule.

Citric acid is produced by fermentation of beet or cane sugar molasses or of glucose. Citric acid is recovered from the fermentation broth in three ways: by the precipitation process, by liquid extraction and by a chromatographic separation process using ion exchange resins or adsorbents.

In the precipitation process, by addition of an equivalent amount of calcium carbonate (Heding and Gupta, 1975) or lime, or $CaCl_2$, citric acid is separated from the fermentation broth as the sparingly soluble calcium salt[14]. The calcium citrate thus obtained is filtered and washed and then reacted with a small excess of H_2SO_4 where calcium precipitates as $CaSO_4$ and filtered off (Pazouki and Panda, 1998). The filtrate contains the crude citric acid which needs to be purified in order to meet the purity standards for food applications. The purification consists of decolorization and demineralization by ion exchange, followed by concentration and crystallization (figure 9.1).

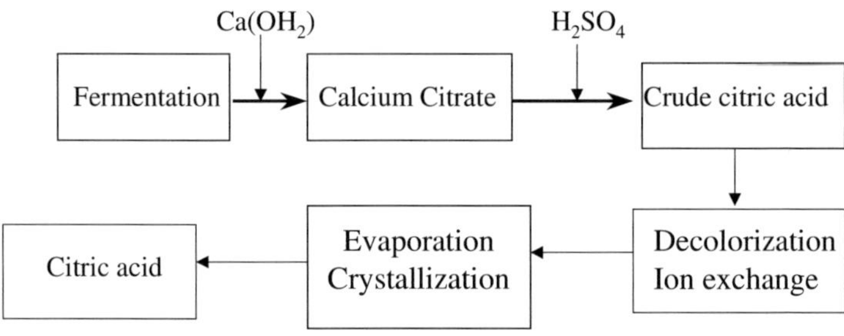

Figure 9.1 Precipitation process for citric acid recovery

The main impurities of the crude citric acid solution that needs to be decolorized and demineralized are salts mostly Ca^{2+} and SO_4^{2-} at about 50 meq/L concentration, free H_2SO_4 and color.
For demineralization, a macroreticular SAC resin in the H^+ form and a WBA resin in the free base form are used. For the decolorization, a synthetic adsorbent can be used or granular activated carbon (GAC). The position of the GAC is before or after the ion exchange, or both, depending upon the color

---

[14] Solubility of calcium citrate: 0.85 g/L at 18°C

230

materials to remove. A synthetic adsorbent can also be used, frequently placed between the SAC and the WBA resins. Activated carbon can also be included:

$$SAC \rightarrow Adsorbent \rightarrow WBA \rightarrow GAC$$

Styrene-DVB adsorbents are used but formophenolic ones can also be used. Formophenolic adsorbents have good decolorization efficiency and also remove other off-taste and odor molecules. The special feature to consider when using formophenolic adsorbents have already been discussed (pages 78-79). Specific flow rate for the adsorbent should be 1 BV/h while for the WBA resin 2 BV/h. Operating temperature about 40°C. Regeneration of the SAC is done with 80 g HCl/L_R counter-current (or 100 g/L_R co-current) using 8% HCl at 1-2 BV/h and 80 g $NaOH/L_R$ for the WBA resin using 4% NaOH and 2 BV/h.

The SAC resin is expected to have an operating capacity of about 1 eq/L_R while the WBA about 1.2 eq/L_R. For that reason, in order to equilibrate the cycle length of the two resins, the SAC resin volume is about 20% more than the WBA resin.

All types of WBA resins can be envisaged here, styrenic, acrylic or formophenolic. Each one has its advantages and disadvantages. The formophenolic resins have high initial operating capacity for SO_4^{2-} removal, they have good decolorization capacity and also they have lower affinity for citrate ions compared to SO_4^{2-} so that at the end of the loading cycle there is less citric acid loaded on the resin. This citric acid will be lost during the regeneration step. In addition, swelling during the loading cycle is less compared to styrenic or acrylic resins and therefore the physical stability is good. Their disadvantage is that they lose capacity and they develop long rinse with time faster than styrene or acrylic resins especially if working temperature is higher than 40°C. The styrenic resins have to be stable physically because, as mentioned earlier (page128) they swell considerably going from the regenerated to the citrate form. However, the styrenic WBA resins that exist in the market do not have all the same characteristics, for example they do not swell to the same extent in citric or lactic acids. Acrylic WBA resins have high physical stability but have long rinse requirements even in new condition.

In a patent assigned to UOP (Universal Oil Products), another mean of separating citric acid from the other constituents of the fermentation broth is presented where citric acid is recovered directly from the fermentation broth using adsorption chromatography (Kulprathipanja, 1986). In this approach, simulated moving bed or a single column was used to fix citric acid on a polymeric adsorbent without any functional groups, with water or a 90% water-10% acetone solution as eluent. The fermentation broth had the composition:

Citric acid	> 13%
Salts	6 g/L
Carbohydrates	1%
Proteins and aminoacids	5%

Polymeric adsorbents of styrene-DVB or acrylic matrix were tested at various temperatures and citric acid concentrations. It was found that the polymeric adsorbent showed a higher affinity for citric acid compared to the salts and the other components of the broth provided that the pH of the feed solution was lower than the first pKa of citric acid. Temperature was 45°C up to

93°C. Water-acetone mixture was used as eluent in order to keep the elution temperature low (45°C).

In a more recent patent assigned to UOP (Universal Oil Products), anion exchange resins were used to recover citric acid and other acids such as lactic and tartaric acids, from fermentation broths using SMB system (Kulprathipanja and Oroskar, 1991). Citric acid then goes out from the extract port while the salts, carbohydrates and proteins go out from the raffinate port.

In the UOP patent the ion exchange resin works in the sulfate form (or SO_4^{2-}/HSO_4^- form). As long as the feed solution had a pH lower than the first pKa of citric acid, the resin had higher affinity for citric acid than for the rest of the components of the feed solution. Elution is performed with low concentration H_2SO_4 (0.01 to 0.2 N H_2SO_4) or with water. It was found that water elution gave a longer retention time for the citric acid peak.

Similar approach was followed more recently using a novel resin made especially for this application in a SMB system (Wu, 2009). The resin was a polyvinylpyridine type containing also carboxylic acid groups.

Applexion® SSMB technology was used (Théoleyre and Baudouin, 2007) to purify citric acid by elution chromatography. The fermentation broth was first centrifuged and filtered with membranes. The pH of the obtained solution was below the pK of the citric acid, at about 0.5 to 2. The resin was an acrylic WBA exchanger having preferably an average particle size of 350-450 μ and a uniformity coefficient of < 1.2. The eluent was dilute H_2SO_4.

In all the above chromatographic purification processes, the mechanism is ion exclusion. Citric acid being neutral molecule at the pH<pKa1, is not excluded by the Donnan potential. In adition, it may form H-bonds with the bisulfate ions fixed by the WBA resin.

Lactic acid

Lactic acid is a monoprotic acid with the formula CH_3-$CH(OH)$-$COOH$. The pKa value is 3.86. It is produced by fermentation using sucrose, lactose (from whey), glucose or maltose from starch.The raw material used will affect the purification process and the ion exchange layout to use.

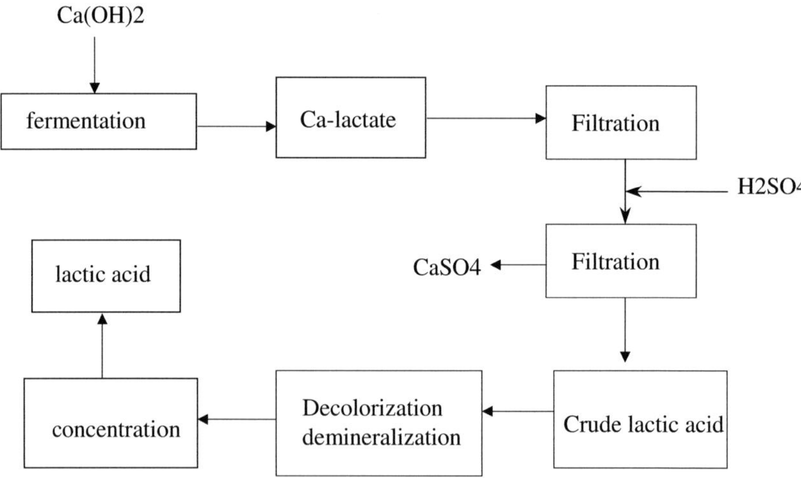

Figure 9.2 Precipitation process for lactic acid recovery

Lactic acid is recovered from the fermentation broth in a similar way as citric acid. The flow sheet for lactic acid recovery with the precipitation process is not much different from that of citric acid (fig. 9.2)

A neutralization agent, $Ca(OH)_2$ or $CaCO_3$ is added in the fermentation broth either from the begining or intermittently. Then the broth is heated and the pH brought to 10-11 where calcium lactate solubilizes. The broth is then filtered to remove cells and other insoluble material. Calcium lactate is then acidified with H_2SO_4 to recover lactic acid. The $CaSO_4$ precipitate is filtered off. The crude lactic acid is then demineralized and decolorized with ion exchange, adsorbent and activated carbon, depending on the raw material used and the grade of lactic acid produced, food or pharmaceutical. If for example raw material is molasses then the lay out used could be:

$$SAC \rightarrow Adsorbent \rightarrow WBA \rightarrow GAC$$

Or if refined sucrose was used:

$$SAC \rightarrow WBA (\rightarrow GAC)$$

The main ionic impurities are Ca^{2+} and SO_4^{2-}. The SAC is a gel or macroreticular type. The WBA resin choice is based on similar considerations as discussed above in the citric acid demineralization. The ion exchange resins are the same as those discussed for the citric acid decolorization and demineralization.

According to the UOP patent discussed above (Kulprathipanja and Oroskar, 1991), lactic acid can also be recovered with adsorption chromatography using the SMB technology, in a similar way as citric acid.

Malic acid

Malic acid is found in apples and many other fruits and vegetables and is used as food acidulant. It is produced from maleic anhydride by double hydration, or from salts of fumaric acid by enzymatic reaction to form the corresponding salt of L-malic acid. When produced from maleic anhydride, the resulting reaction mixture contains, except of the DL-malic acid, impurities such as maleic acid, fumaric acid, color materials and eventually metal ions from the reactors. To obtain pure DL-malic acid, the reaction mixture passes through an anion exchanger, either weak base or strong base (Sumikawa *et al*, 1976).

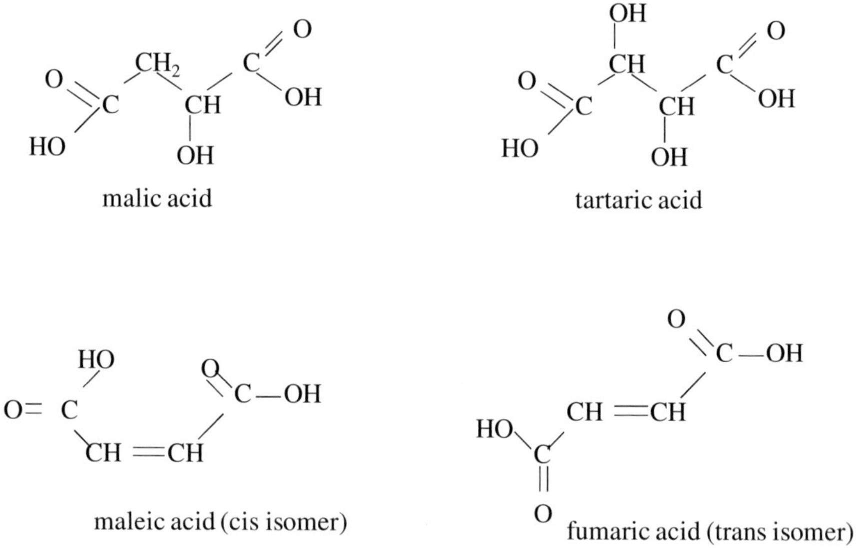

Figure 9.3 Chemical formula of malic, maleic, fumaric and tartaric acids

Maleic acid is a relatively strong acid compared to malic and fumaric acids. pKa$_1$ for maleic, malic and fumaric acids is 1.9, 3.46 and 3.03 respectively. The anion exchangers have a higher affinity for the stronger acids and lower selectivity for the weaker acids. When the mixture passes through the resin, maleic acid is retained stronger and displaces the other two acids which leak first. The cycle stops when maleic acid starts leaking through. Fumaric acid was found to be removed from the effluent with activated carbon which in addition removes color matter. The effluent from the activated carbon column passes finally through a SAC resin in H$^+$ form which removes any metal cations present.

The reaction mixture obtained from the enzymatic reaction for producing L-malic acid from fumaric acid salts contains malic acid and unreacted fumaric acid in the salt form. In order to recover pure L-malic acid, the mixture passes through a SAC resin in the H$^+$ form where the cations are removed from the solution and the effluent contains L-malic and fumaric acids. The mixture of acids is then concentrated by distillation at 80°C at reduced pressure until malic acid is found at a concentration of about 50-60%. By cooling down to 10-15°C fumaric acid precipitates out. The solution is filtered to recover pure L-malic acid. The SAC resin is subsequently regenerated with HCl or sulfuric acid. The fumaric acid precipitate is converted to the salt form and recycled back in the process.

Succinic acid

Succinic acid is a naturally occurring dicarboxylic acid having the formula $HOOC\text{-}CH_2\text{-}CH_2\text{-}COOH$. It is used as flavouring agent in food and beverages. It can be produced by catalytic hydrogenation of maleic acid or its anhydride. More recently, fermentation manufacturing processes have been developed using carbohydrates as raw material.

The process to recover succinic acid from the fermentation broth is similar to that described for citric and lactic acid (Datta *et al*, 1992). Succinic acid is precipitated from the fermentation broth by adding calcium hydroxide or calcium oxide where the insoluble calcium succinate is formed. The precipitate is removed by filtration and it is used to form a slurry to which H_2SO_4 is added where insoluble $CaSO_4$ is formed along with soluble succinic acid.

The reaction mixture is filtered to remove the CaSO4 precipitate and the solution of succinic acid is purified with ion exchange.

The ion exchange system used is a SAC followed by a WBA exchange resin.

In another process (Gerberding and Singh, 2012) succinic acid is produced by fermentation as ammonium succinate. Succinic acid is obtained either by passing the ammonium salt through a SAC resin in the H^+ form, or through a SBA resin where the succinate anion is fixed and the NH_4^+ leak through. Previously, the broth is purified by treating with micro- and UF membranes. If a SAC resin is used, the NH_4^+ is regenerated from the resin using H_2SO_4. If a SBA resin is used, the anion resin is eluted with a strong acid such as H_2SO_4 or HCl, which elutes quantitatively the succinic acid being a weak acid.

238

10. Gelatin purification

Collagen is a group of proteins found in animals. It is mostly found in fibrous tissues such as tendons, ligaments and skins, also in cartilage and bones. It is a protein that has a negative net charge and shows a certain cation capacity. Gelatin is collagen that has been irreversibly hydrolyzed. Pure gelatin obtained from lime conditioned stocks has weakly acidic properties. Therefore, at its isoelectric point gelatin contains other, external, ions so that the net charge is zero. For example, if HCl is added to pure gelatin, the HCl will partially suppress the ionization of the carboxylic acid groups while it will partially convert the amine groups to the ammonium form. The point where the negative charges balance the positive charges is the isoelectric point. If all external salts are removed, by utilising for example a mixed bed of a SAC and a SBA exchange resin to remove all non-gelatin cations and anions, then gelatin reaches an isoionic point[15]. Isoelectric and isoionic points are expressed in pH units. In terms of pH, gelatin has an isoelectric point somewhere between pH 4.7 and 9. Gelatin forms a viscous solution with

---

[15] Isoionic point is the pH value at which a zwitterion has equal number of positive and negative charges, without any adherent ionic species, that is, besides H^+ and OH^-, no other ions are present. Isoelectric is the pH value at which the net charge of the molecule, including bound ions, is zero.

water when heated which upon cooling it gels. Important properties of gelatin are viscosity and its ability to form thermoreversible gels characterized by the gel strength.

The main raw materials for gelatin production are bones and skin. Bones are first treated with dilute acids to remove calcium salts and hot water or solvents are used for degreasing. Hides are washed to remove hair and small particles and then they are degreased. The raw materials are then prepared for extraction either with acids or with alkali treatments. Gelatin made from acid treatment is called type A gelatin and has an isoionic point 7 to 9 while that made from alkali treatment is called type B gelatin and has isoionic point 4.8 to 5.2.

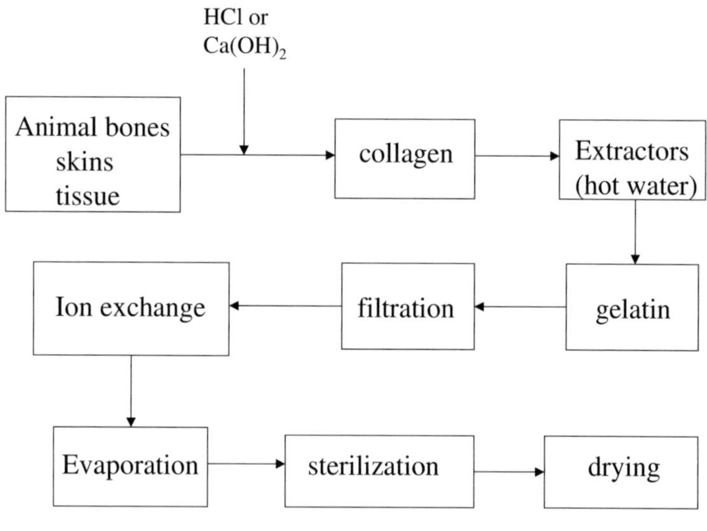

Figure 10.1

The partially purified collagen goes to the extractors where gelatin is extracted with water or acid solutions (figure 10.1). The solution that comes out of the extractors has a gelatin concentration of about 5%. It is pre-purified by filtration to remove any fats, grease or insoluble particles, after which it is treated by ion exchange resins to remove the salts that were formed during the acid or alkali treatments and washings.

The gelatin solution is demineralized with SAC resin in the H^+ form followed by WBA and sometimes by a SBA resin in the OH^- form:

$$SAC\ (H^+) \rightarrow WBA\ (\rightarrow SBA\ (OH^-))$$

The SAC resin is of a gel type. WBA resins are of gel type acrylic or of MR type styrenic. The SBA resin is gel type styrenic. The feed solution contains 5% gelatin, mineral salts and free acids depending on the gelatin type and treatment before the IER. Chlorides, sulphates and phosphates from the bones and the acids are present along with sodium, potassium and calcium. The relative resin volumes of SAC and WBA depend on the feed solution composition. Regeneration levels are 70-80 g HCl/L_R for the cation and 100 g $NaOH/L_R$ for the anions. Service cycle flow rates are in the range of 4-8 BV/h and the temperature is around 50-70°C to keep viscosity and linear velocity of the feed solution compatible with pressure drop restrictions. In fact, pressure drop and capacity reduction may be a problem due to resin fouling from deposition of fats and peptides on the surface and inside the resin beads. In the regeneration steps, a backwash is included to remove any substances that may plug the resin column.

Sometimes, backwashing with warm brine may clean the resins from these materials. Clean up with 10% HCl solution applied for a few hours may also be effective in cleaning the resin. Sometimes the resin is fouled with iron originating from haemoglobin in the raw stock. In this case, treatment with mild reducing agents may reduce Fe^{3+} to Fe^{2+} and remove it from the resin with a brine wash.

11. Glycerin purification

Glycerol is a simple polyol compound, $HOCH_2\text{-}CHOH\text{-}CH_2OH$, widely used in pharmaceutical formulations and in food and beverages as humectant, solvent and sweetener. The term glycerin applies to commercial grades of glycerol containing different glycerol levels. It can be obtained from natural fats and oils via three routes:

- in soap manufacture
- in fatty acids manufacture
- in transesterification to obtain biodiesel

In soap manufacture, fat is boiled with caustic soda and salt solution whereby it is formed soap and glycerol.

$$
\begin{array}{l}
CH_2OCOR_1 \\
|\\
CHOCOR_2 \;+\; NaOH \;\rightarrow\;
\end{array}
\qquad
\begin{array}{l}
R_1COONa \\
R_2COONa \;+\; \\
R_3COONa
\end{array}
\qquad
\begin{array}{l}
CH_2OH \\
|\\
CHOH \\
|\\
CH_2OH
\end{array}
$$

Oils, fats   soaps   glycerol

A separation into two phases takes place, the upper layer being the soap and the lower layer (called spent lye) containing glycerol, water, salt and excess caustic. However, the spent-lye crude contains up to 10% ash and this makes the use of ion exchange uneconomical.

In fatty acids manufacture, the fat is hydrolyzed at high pressure and temperature whereby the water splits the fat into fatty acids and glycerol.

$$
\begin{array}{c}
CH_2OCOR_1 \\
| \\
CHOCOR_2 \\
| \\
CH_2OCOR_3
\end{array}
\;+\; H_2O \;\xrightarrow{\;\Delta,\,P\;}\;
\begin{array}{c}
R_1COOH \\
R_2COOH \\
R_3COOH
\end{array}
\;+\;
\begin{array}{c}
CH_2OH \\
| \\
CHOH \\
| \\
CH_2OH
\end{array}
$$

Oils, fats   Fatty acids   glycerol

The fatty acids are removed leaving at the bottom of the column the glycerol containing aqueous phase called sweet water. Hydrolysis crudes contain <1% ash and <1.5% organics. Ion exchange in this case is a purification process used industrially.

Industrial processes to purify hydrolysis crude glycerol are based on the following flow sheet:

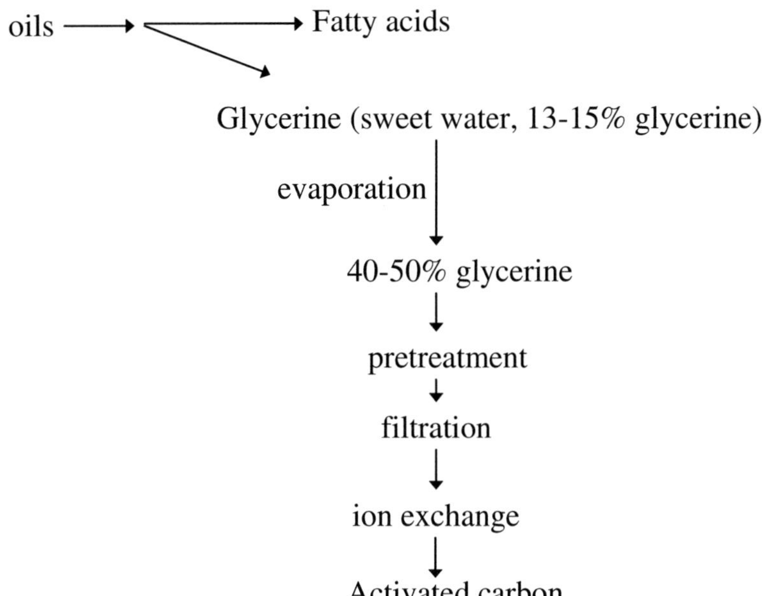

The pretreatment may involve an addition of lime which converts any fatty acids to insoluble compounds. Then by adding Na_2CO_3 the remaining lime is precipitated as $CaCO_3$ (Birbingham 1967). All these precipitates are removed by filtration before entering the ion exchage plant.

The ion exchange process consists of all or some of the following ion exchange resins:

WAC → SAC →adsorbent → WBA → SBA

This layout can constitute a first step, followed by a second step:

SAC $\rightarrow$ SBA $\rightarrow$ (SAC/SBA) mixed bed

The first step removes the bulk of the mineral impurities while the second step upgrades the quality of the glycerol. Typical operating conditions are: T=60-65°C, pH=7-9, glycerine concentration 40-50%, specific flow rate 2-6 BV/h, regeneration: cation resins, 80-120 g HCl/L_R, 60-100 g $NaOH/L_R$ (it depends on the type of cation and anion resins used, that is, the proportions weak and strong electrolyte resins, and on the ionic impurities, Na^+, Ca^{2+} etc).

In transesterification, usually with methanol, to make biodiesel, the fat reacts with methanol to form methyl esters and glycerol, which is separated from the methyl esters by water wash.

$$
\begin{array}{ccccc}
CH_2OCOR_1 & & & R_1COOCH_3 & CH_2OH \\
| & & & & | \\
CHOCOR_2 & + CH_3OH & \xrightarrow{catalyst} & R_2COOCH_3 & + CHOH \\
| & & & & | \\
CH_2OCOR_3 & & & R_3COOCH_3 & CH_2OH \\
\end{array}
$$

Oils, fats   Fatty acids esters   glycerol

Most frequently the reaction uses homogeneous alkaline catalysts such as sodium methylate, potassium methylate, NaOH or KOH. The glycerin phase is neutralized with acids and the formed salts are found, along with the other impurities like soaps, methanol, water and catalyst, in the glycerin phase. Crude glycerin from transesterification is a major by-product in the production of biodiesel. It is usually of better quality than from other routes but still it contains a few to several percent of salts. Glycerin as a biodiesel by-product, represented a major source of crude glycerin and its purification representedan important parameter for theexisting or new applications. However, with today's decrease in biodiesel production, this process lost interest.

The crude glycerin from biodiesel is first treated with acid to split the soaps formed from the fats and to remove the alkaline catalyst (Hajek and Skopal, 2010). The acid treatment produces free fatty acids (FFA) and salts. Since the FFA are not miscible with glycerine they float at the top and can be skimmed off. Some of the salts precipitate out. Following this pretreatment, methanol is removed by flash evaporation. The remaining glycerine has a purity of about 85% and is further treated with caustic to remove any remaining fats and with activated carbon to remove any colors and odors. Then ion exchange and adsorption follows. Conventional ion exchange processes as the ones described above are uneconomical for the level of impurities contained in crude glycerin from biodiesel. Here, ion exclusion technology is feasible (Prielipp and Keller, 1955; Asher and Simpson, 1956; Klipper*et al*, 2007) and if necessary it can be followed by a conventional ion exchange polishing unit like a mixed bed for bringing the salts level down to parts per million. Ion exclusion chromatography can be combined with SMB or similar technologies (Lancrenon and Fedders, 2008).

12. Nutraceuticals

Some food ingredients have been found to provide health benefits like decreasing the risk for certain diseases or helping healing others. These food ingredients are called nutraceuticals (from the words **Nutr**ition and Pharm**aceuticals**), and include vitamins, aminoacids and polyphenolic compounds. Nutraceuticals may be taken in normal eating by well choosingthe food, or can be taken as dietary supplements, which are taken apart from food as tablets, liquids, pills etc, or as fortified foods such as milk enriched in vitamin D. Nutraceuticals can be prepared via a synthetic route, fermentation or from extracts from animals or plants. The use of IER in aminoacids processing has already been discussed in chapter 8. Here are discussed vitamins and polyphenols.

Vitamins

In 1910, the Japanese scientist Umetaro Suzuki (1874-1943) discovered a substance that he patended as *aberic acid*. Later this compound was identified as thiamine. In fact, after the discovery of aberic acid, the Polish biochemist Casimir Funk proposed the term Vitamine from the latin word Vita (=life) and amine *(vital amine)*. When it was found that not all vitamins contained an amine group, the "e" was dropped from Vitamin-e to differentiate clearly from the amines, and the term Vitamin is now used.

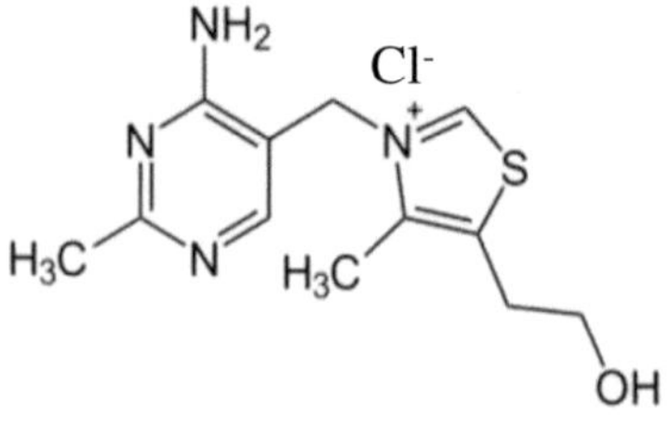

Figure 12.1 Thiamine (vitamin B_1)

As indicated by its chemical structure, thiamine can be fixed on a SAC exchanger by the quaternary ammonium group. In one process for example, Thiamine is made by fermentation, and subsequent recovery from the fermentation liquor by passing through an ion exchange resin. Vitamin B1 is finally recovered by elution from the ion exchange resin and precipitation from the eluate (Silhankova, 1977).

Vitamin B_{12} has been extracted from fermentation broths with weak acid cation exchangers of acrylic acid-DVB type (Lightfoot, 1961). Vitamin B_{12} is loaded on the resin at a pH of 2.5-3.0, where the carboxylic groups of the ion exchanger are not dissociated. Since vitamin B_{12} is essentially non-ionic, an absorption mechanism may be involved. The high resin selectivity for B_{12} may also suggest this kind of mechanism. Elution of vitamin B_{12} from the resin was achieved with NaOH, KOH or NH_4OH at a pH between 7 and 9. Furtheremore, weak acid cation exchange resins have been used as stabilizing agents for vitamin B_{12}. In fact, some materials are harmful to vitamin

250

B_{12} such as ascorbic acid. In addition, it appears that vitamin B_{12} is partially destroyed by the gastric juice. The resin-complex form of vitamin B_{12} is stable in the stomach and the vitamin will not be eluted from the resin until it enters the intestine (Bouchard et al, 1958).

Vitamin C is used as vitamin supplement in food and beverages as well as antioxidant in canned and frozen prepared foods. L-ascorbic acid is the recognized name for Vitamin C.

The most important chemical characteristic of L-ascorbic acid is its reversible oxidation to monodehydro- L-ascorbic acid and dehydro-L-ascorbic acid, which is the basis for its physiological activity as antioxidant and radical scavenger.

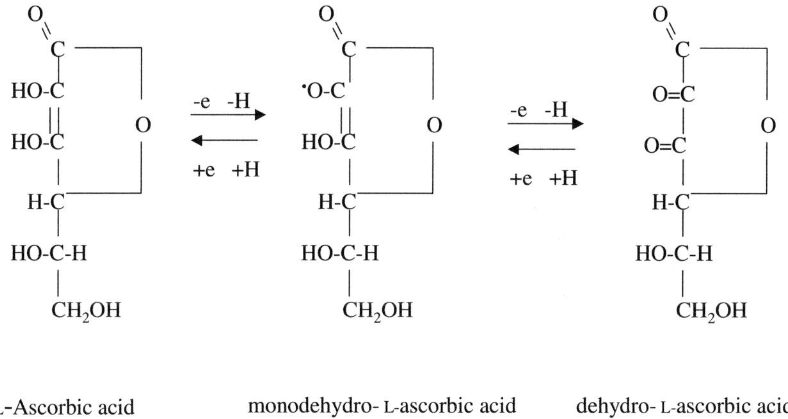

L-Ascorbic acid   monodehydro- L-ascorbic acid   dehydro- L-ascorbic acid
(free radical)

Figure 12.2

In fact, vitamin C is a redox system containing the above three substances. Ascorbic acid and its sodium, potassium and calcium salts are used as food antioxidants. These compounds are water soluble and therefore they cannot protect fats and oils from oxidation. For that, the fat-soluble ascorbyl palmitate or ascorbyl stearate can be used as food antioxidants.

The raw material for the synthesis of L-ascorbic acid is D-glucose (Kuellmer, 2001). At first, D-glucose is catalytically hydrogenated to D-sorbitol followed by microbiological oxidation to L-sorbose (figure 12.3).

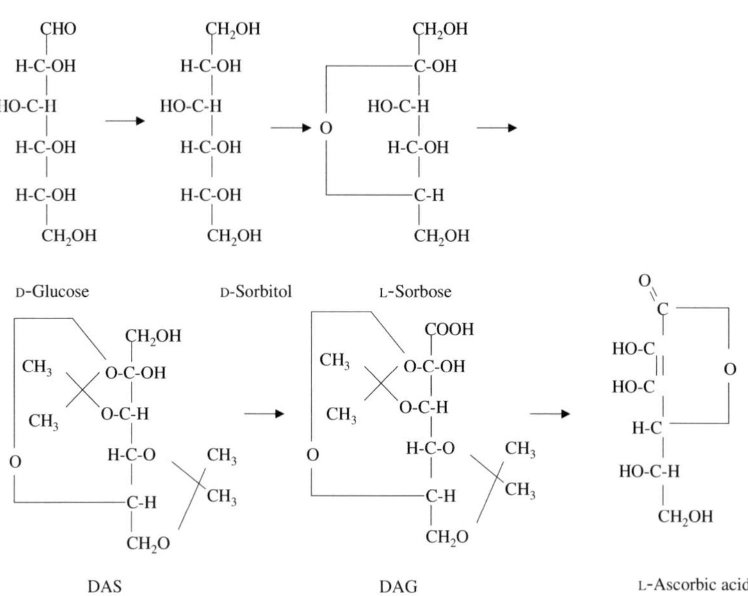

Figure 12.3 Chemical synthesis of vitamin C

L-Sorbose is reacted with acetone and sulfuric acid to yield 2,3:4,6 diacetone-L-sorbose, (DAS). The remaining primary hydroxyl group is oxidized to carboxylic acid to produce 2,3:4,6-diacetone-2-keto-L-gulonic acid (DAG) which is subsequently treated with hydrochloric acid to form L-ascorbic acid.

A variant of this synthetic route is the oxidation of L-sorbose by fermentation directly to 2-ketogulonic acid (2-KGA) (figure 12.4). 2-KGA is then converted to ascorbic acid by acid conversion. A different route is to convert the intermediate 2-KGA to an alkyl ester from which ascorbic acid can be obtained by refluxing in an organic solvent in the presence of a base such as sodium bicarbonate or sodium carbonate (Wang *et al*, 2012). Ascorbic acid is obtained as the sodium salt which precipitates out on cooling. The crude salts are recovered by filtration, mixed with water from which ascorbic acid can be obtained using a SAC resin in the H^+ form. Ascorbic acid is obtained by crystallization from a methanol-water solution.

A process for purification of ascorbic acid from other compounds like 2-KGA and salts involves a chromatographic separation using preferably SMB technology for a more efficient separation (Kawai *et al*, 1998). The strong acid cation exchange resin used in this process works in an ionic form corresponding to the mineral impurities of the feed solution, for example mixed H^+ and Na^+ form. Water is used as eluent. Ascorbic acid is the slow moving compound and comes out in the extract port. Impurities come out in the raffinate port.

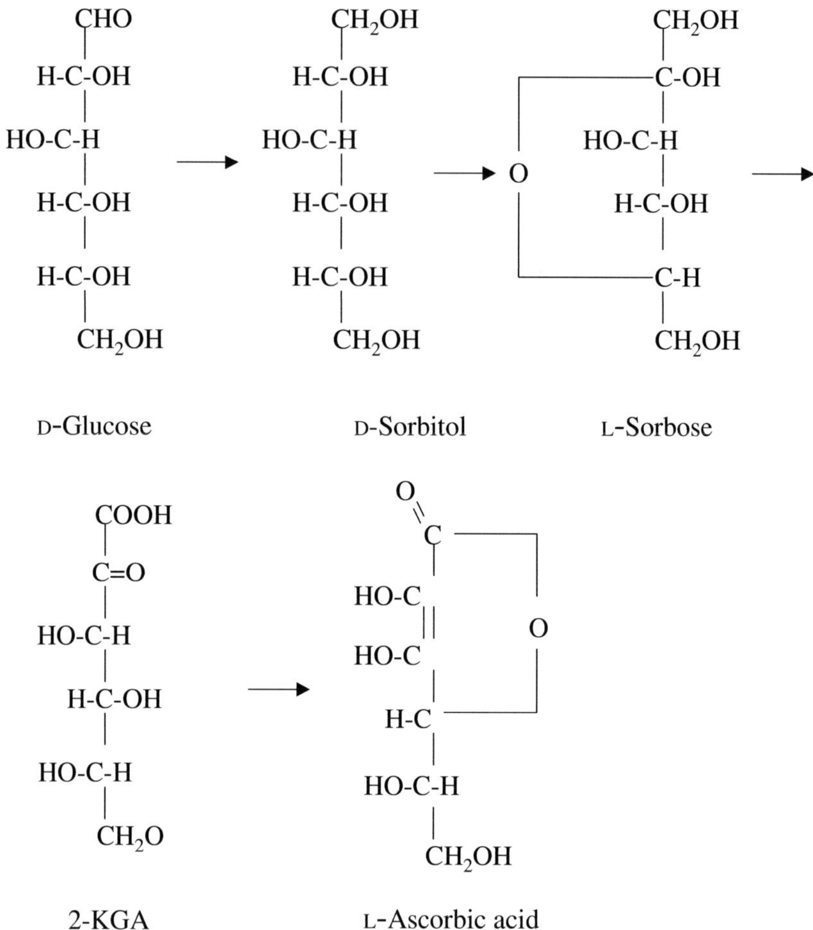

Figure 12.4 Combined fermentation and chemical synthesis of ascorbic acid

254

In another patent (Arumugam *et al*, 2003) the suggested continuous process starting from 2-KGA (fig. 12.5) comprises the following steps:

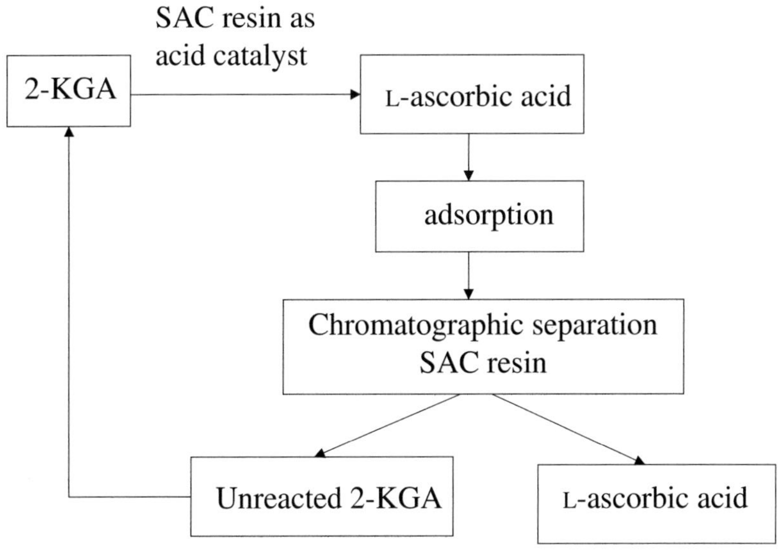

Figure 12.5 continuous process for ascorbic acid production
(Arumugam *et al*, 2003)

The acid-catalyzed conversion of 2-KGA to ascorbic acid was achieved with a sulfonic SAC resin used as a catalyst. Before the chromatographic separation of ascorbic acid from unreacted 2-KGA, a synthetic adsorbent (Dowex® L-285) or activated carbon was used to remove high molecular weight colored material. The chromatographic separation of 2-KGA from ascorbic acid was achieved with a gel type SAC chromatographic resin of narrow particle size distribution and an average diameter of 320 µ using SMB technology. Water is used

255

as eluent. The unreacted 2-KGA fraction (from the raffinate port) was recycled back to the reactor to be converted to ascorbic acid. Ascorbic acid from the extract port had better than 90% purity and it was then crystallized to isolate the ascorbic acid.

The chromatographic separation of ascorbic acid and 2-KGA is based on the ion exclusion principle and the acidity difference between these two compounds: the pKa of ascorbic acid is 4.2 while 2-KGA is a stronger acid with a pKa of 2.66. From the mixture of the two compounds, the more dissociated 2-KGA is excluded while ascorbic acid has a longer retention time.

Vitamin E is a group of compounds that include the tocopherols and the tocotrienols (figure 12.6). It is a fat-soluble vitamin with antioxidant properties that stops the production of reactive oxygen species formed in the oxidation of fats. Of the tocopherols and the tocotrienols, the α-tocopherol is the most biologically active form. Most of the dietary supplements of vitamin E provide only α-tocopherol. Frequently, α-tocopherol in dietary supplements and fortified foods is esterified (α-tocopheryl acetate and succinate) to extend its shelf life. The body hydrolyzes these esters to use the antioxidant properties of vitamin E.

Vitamin E is prepared from trimethylhydroquinone (TMHQ) and isophytol (IP) (figure 12.6) and therefore the synthesis of TMHQ and IP are the first steps for the synthesis of vitamin E. IP is synthesized starting from acetone, ethyne and hydrogen through a combination of C2 and C3 elongation reactions (Bonrath *et al*, 2005). In one of the pathways, for the ethynylation of ketones and aldehydes, gel type strong base anion exchange resins such as Amberlite® IRA400 of Rohm and

Haas Co (now The Dow Chemical Company) are used as heterogenious catalysts.

tocopherols

R$_1$	R$_2$	R$_3$	name
CH$_3$	CH$_3$	CH$_3$	α-tocopherol
CH$_3$	H	CH$_3$	β-tocopherol
H	CH$_3$	CH$_3$	γ-tocopherol
H	H	CH$_3$	δ-tocopherol

tocotrienols

R$_1$	R$_2$	R$_3$	name
CH$_3$	CH$_3$	CH$_3$	α-tocotrienol
CH$_3$	H	CH$_3$	β-tocotrienol
H	CH$_3$	CH$_3$	γ-tocotrienol
H	H	CH$_3$	δ-tocotrienol

Figure 12.6 Vitamin E

The last step in the synthesis of α-tocopherol is the acid-catalyzed condensation reaction of IP with TMHQ (figure 12.6). Various acid catalysts or systems of catalysts have been described. Macroreticular type SAC resins in H$^+$ form, such as Amberlyst$^®$ 15 of Rohm and Haas Co (nowThe Dow Chemical Company) have been used as heterogeneous catalysts (Moroe*et al*, 1969) to produce α-tocopherol. The reaction takes place in an inert solvent such as toluene or benzene at the temperature where an azeotrope is formed between the solvent employed and water. The formed water during the reaction (fig. 12.7) is then azeotropically removed from the reaction medium.
The reaction medium is separated from the catalyst resin by simple filtration. In addition, the formed α-tocopherol can

subsequently be converted to an ester with, for example, acetic anhydride or succinic anhydride, using the same IER as catalyst.

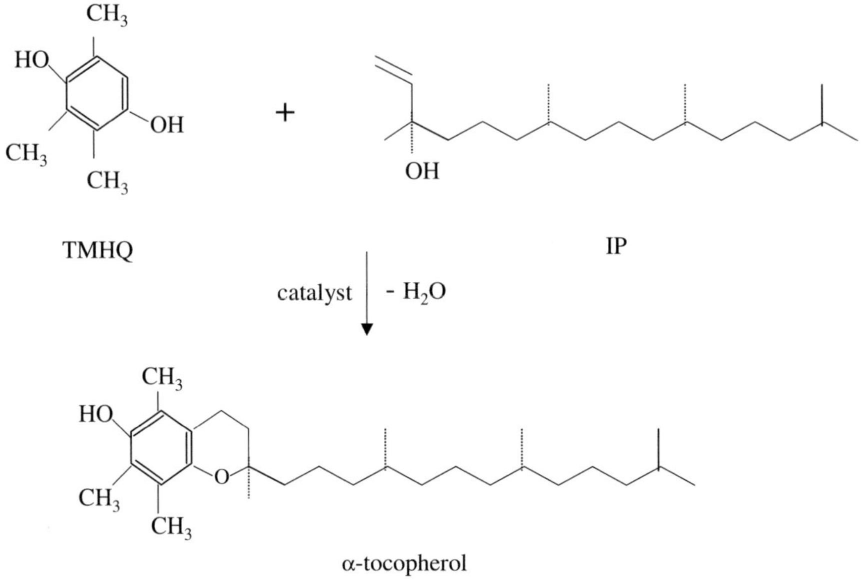

Figure 12.7 synthesis of α-tocopherol

Ion exchangers of the perfluoroalkylenesulfonic acid type such as Nafion®NR50 of E.I.Du Pont de Nemours & Co., Inc., known as "superacids", have preferably been used as heterogeneous catalysts for the above reaction (Bonrath, 1999).

Non-alpha-tocopherols (fig. 12.6) can be converted to a-tocopherol by reacting with formaldehyde using a strong acid ion exchange resin such as Nafion®NR50 of E.I.Du Pont de

Nemours & Co., Inc.,due to high temperatures required, and a hydrogenation catalyst (Nelan and Foster, 1980).

Tocopherols and tocotrienols have been recovered from palm fatty acid distillates using ion exchange resins (Top *et al*, 1993). The method consisted in reacting first the fatty acids and glycerides with an alkyl alcohol in presence of a catalyst to convert them to alkyl esters which are subsequently removed by distillation leaving the tocopherols and tocotrienols in the residue. After filtration to remove any precipitates, the tocopherols and tocotrienols were fixed on an anion exchange resin in the OH⁻ form. Elution was performed with 10% H_2SO_4 and the tocopherols and tocotrienols were collected with 96% ethanol. The resin was regenerated with NaOH for the next cycle. Similarly, a strong base anion exchanger was used to purify natural vitamin E concentrate (Yu*et al*, 2006). With 52% vitamin E concentration in the feed ethanol solution, the resin was reported to have a saturation capacity of 630 mg/g dry resin at 25°C which means about 1.5 meq/g dry resin, that is around 35% of the total exchange capacity of a typical SBA resin.

Polyphenols

Phenolic compounds are defined as those substances that have one or more hydroxyl groups bound on an aromatic ring. Compounds that have many phenolic hydroxyl groups are often called polyphenols. Due to their chemical structure, phenolic compounds have radical scavenging and antioxidant activity

(Maestri *et al*, 2006). Phenolics are found in many fruits and vegetables. They are responsible for organoleptic properties such as color and taste but many studies carried out in vivo and in vitro have indicated that phenolic compounds contribute to health benefits (Gattuso *et al*, 2007). For example, foods and beverages rich in phenolic compounds have been associated with a decreased risk of developing cardiovascular diseases and some cancers, property that has been attributed to their strong antioxidant activities. The evidence for protective properties of polyphenols for various diseases has increased the development of polyphenol-rich foods and supplements.

Polyphenols comprise a great variety of compounds, from simple molecules to polymers. They are classified in flavonoids and non-flavonoids. Non-flavonoids are phenolic acids and stilbenes but also complex molecules derived from them, like gallotannins, ellagitannins and lignins. Flavonoids share a common nucleus (figure 12.8) and depending on the oxidation state of the heterocyclic pyran ring, they include flavonols, flavanols (catechins) and anthocyanidins (figure 12.9).

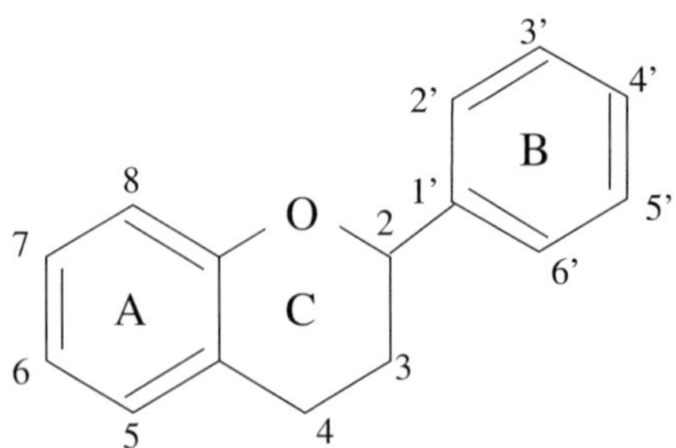

Figure 12.8 Chemical structure of the flavonoids backbone

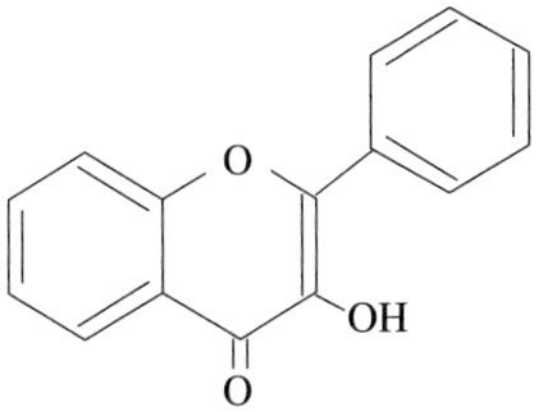

flavonols

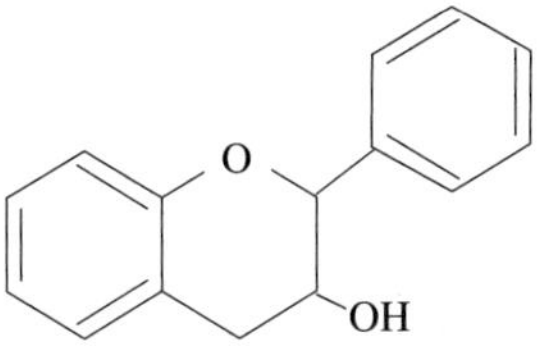

flavanols

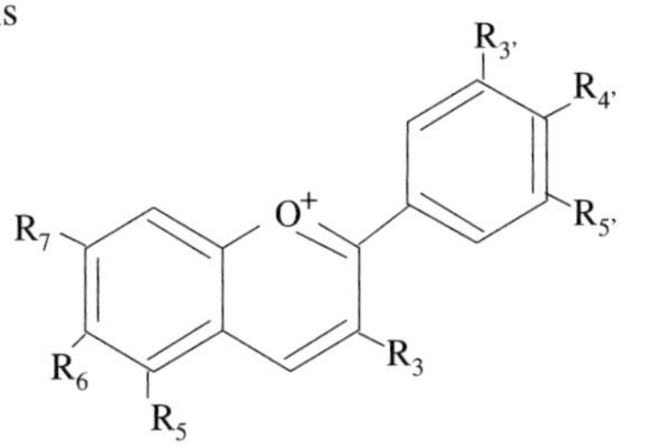

anthocyanidins

Figure 12.9 Chemical structure of flavonols, flavanols and anthocyanidins

Polyphenols can be recovered from plants and fruit products and by-products. The basic approach for the recovery of polyphenols from plants described in what follows consists of three steps: extraction of the phenolic compounds from the raw material, concentration on a synthetic adsorbent and recovery from the adsorbent by elution. The phenolic compounds are then recovered from the concentrated eluate. In the recent years, much attention has received the recovery of value added compounds used in food from plant wastes (Laufenberg *et al*, 2003). In the Mediterranean countries, olive oil processing wastes represent a major disposal problem but also a source for recovery of valuable phenolic compounds (Fernandez-Bolanos *et al*, 2004). Other wastes becoming sources for polyphenols recovery include hydrolysates from treatment of lignocellulosic residues of agricultural wastes (Parajo *et al*, 2008), grape pomace from vineyards wastes (Mavrakis, 2009) and citrus fruits juices wastes and by-products (Di Mauro *et al*, 2002).

The main phenolic compound in olive oil is hydroxytyrosol (HT) and derivatives (fig. 12.10). HT is believed to be among the most powerful antioxidants.

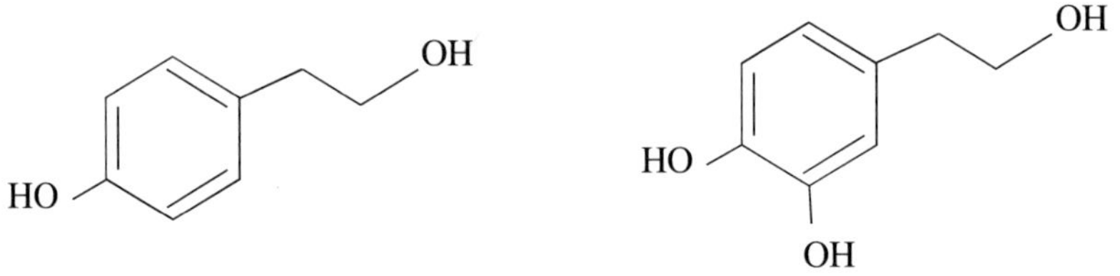

Figure 12.10

During olive oil processing, and depending on the olive oil extraction process, most part of HT is found in the olive mill waste waters (OMWW) for the three-phase process or in wastes consisting of a mixture of solids and liquids in the two-phase process (Fernandez-Bolanos *et al*, 2006). In a patented process (Fernandez-Bolanos *et al*, 2005) HT is isolated from two-phase wastes by first hydrolyzing the solid wastes at high temperatures and acidic conditions, in order to allow HT to be transferred to the liquid phase and be easier recovered. After filtration, the

262

liquid passes through a strong base anion exchanger in the Cl⁻ form where HT is retained by the resin. Elution is performed with water where HT is recovered at a purity of about 60-70 %. The resin is regenerated (cleaned) with a solution of NaOH followed by conversion of the resin to the Cl⁻ form with HCl. The eluate can then pass through a synthetic adsorbent of Amberlite® XAD type made by Rohm and Haas Company (now The Dow Chemical Company) where HT is retained. Elution is performed with a 30% methanol or ethanol-water mixture from where HT is recovered after distilling off the solvent, with a purity of 95%. The adsorbent is subsequently treated with a dilute solution of H_2O_2 followed by a pH adjustment with a NaOH solution. The above two products, with 60-70% purity or with 95% purity have reached commercialization under the names Hytolive® 1 and Hytolive® 2 respectively. Uses of HT include formulations for dressings, tomato products and spreads. In the same context, the same research group investigated the possibility of recovering in addition to HT, other value food compounds such as mannitol and oligosaccharides from the hydrolysates of the two-phase olive oil wastes (Fernandez-Bolanos *et al*, 2004). In this work, mannitol was purified with SAC and SBA resins.

Phenolic compounds recovery from OMWW in three-phase process mills was achieved using synthetic adsorbents (Agalias *et al*, 2007) as follows. The waste waters were first filtered down to 25 μ size filters then passed through a column of Amberlite® XAD16 synthetic styrene-DVB type adsorbent of Rohm and Haas Company. After the loading step, the adsorbent was regenerated with 50/50 ethanol/isopropanol mixture and rinsed with water. The effluent from the Amberlite® XAD16 passed then through a column containing Amberlite® XAD7 synthetic adsorbent with an acrylic backbone made by Rohm and Haas Company. Regeneration of the second column was performed in the same way as the first column. The cycle length was defined so that the first column, containing Amberlite® XAD16, removed the phenolic compounds while the second column, containing Amberlite® XAD7, removed color and odor material and the phenolic compounds that leaked from the first column. At the end of the cycle, 87% of HT and 100% of tyrosol (T) were retained in the first column and the rest 13% of HT was retained in the second column.

The spent regenerants of the two columns were then evaporated to recover and recycle the solvents and to recover the residues containing the eluted phenolic and other compounds. The compounds contained in these residues were then purified and isolated using Fast Centrifuge Partition Chromatography. A 90% pure HT was obtained from the eluate from the adsorbent Amberlite® XAD16.

Overall with this process,
i) the waste waters were deodorized, decolorized and the COD reduced by 98%,
ii) an extract was obtained containing polyphenols and lactones,
iii) an extract was obtained with coloring substances from the olive fruit and
iv) pure HT was recovered.

Citrus fruits contain essentially flavonoids (hesperidine, narirutin, naringin, neohesperidin and anthocyanins in the pigmented oranges). The increased consumption of citrus juices has resulted in an accumulation of wastes and by-products such as seeds, peels, cell and membranes residues, containing significant quantities of polyphenols and a lot of research has focused on the extraction of these polyphenols from citrus peels for preparation of natural antioxidants (or sweeteners from neohesperidine).

Adsorption of model phenolic compounds on several synthetic adsorbents has been studied (Scordino *et al*, 2003 and 2004) by determining the equilibrium loading of the adsorbent as a function of solution concentration at various pH and temperature values. The synthetic adsorbents varied in chemical composition (styrene-DVB, acrylic) and covered a range in specific surface area from 300 to 1200 m^2/g and (average) pore diameter from 20 to 260 Å. Similarly, adsorption of catechin model compound (Gogoi *et al*, 2010) on various synthetic adsorbents and activated carbon was studied by constructing equilibrium isotherms at various pH and temperatures. In this study acrylic type adsorbents showed a higher adsorption capacity compared to styrene-DVB types possibly due to the hydrophilic nature of the adsorbate and the polarity of the acrylic adsorbent.

Removal and recovery of anthocyanins from pulp wash (PW) liquids in citrus processing was studied using model compounds (Scordino *et al*, 2004) and real PW liquids (Di Mauro *et al*, 2002). In the latter study, the PW was treated on a number of adsorbents of which aromatic DVB types were retained, Sepabeads® SP70 and Relite® EXA-90 from Mitsubishi Chemical Corporation, due to their high loading capacities with phenolic compounds. The adsorbed anthocyanins were eluted with 96% ethanol, where all of the phenolic compounds fixed were eluted. In a later study (Scordino *et al*, 2005) elution was looked more closely in order to elute selectively anthocyanins and separated from the rest of the phenolics. The adsorbent used was Relite® EXA-118 from Mitsubishi Chemical Corporation giving higher loading capacities possibly due to its higher specific surface area. Thus, it was found that a 50:50 methanol-water mixture selectively eluted anthocyanins and hydroxycinnamates while with a subsequent treatment with alkaline ethanol solution it was eluted flavanones (narirutin, hesperidin, limonin) (Scordino *et al*, 2006).

Recovery of hesperidin from citrus peel extracts was studied by the same research group as above (Scordino *et al*, 2003). Equilibrium isotherms were constructed using different adsorbents, styrene-DVB or acrylic types. The data were well fitted to the Freundlich isotherm from which the K_F and b_F constants were derived. The constant K_F, which is a measure for the capacity, was then correlated with specific surface areas, S_A, and the pore radius, P_R, of the adsorbents used. It was found that K_F increased as the S_A increased and as P_R increased for $P_R < 90$ Å and above this value, P_R had little effect. The effect of P_R on equilibrium loadings of the adsorbent is possibly related to the molecular size of hesperidin in comparison to the size of the pores of the adsorbent. Based on this study the best adsorbent found among the ones tested was a styrene-DVB with 1200 m^2/g S_A and 90 Å P_R.

Hesperidin was extracted from citrus peels with saturated $Ca(OH)_2$ solution where pectines were precipitated allowing the separation from hesperidin (Di Mauro *et al*, 1999). The clear extracts were neutralized and passed through a styrene-DVB adsorbent to recover hesperidin. Elution was performed with 0.5 N NaOH solution in 10:90 ethanol-water. Hesperidine was precipitated from the concentrated eluate by acidification.

Amberlite® XAD7 acrylic adsorbent from Rohm and Haas Company was used to recover narirutin from water extracts from peels of *Citrus unshiu* (Kim *et al*, 2005). Batch loadings were conducted and the data fit with the Freundlich's isotherm. Elution was performed with 80% methanol, eluting more than 89% of the narirutin fixed on the adsorbent.

Recovery of phenolic compounds from apple juice was studied (Kammerer *et al*, 2007) using an acrylic adsorbent. The study consisted in carrying batch loading and elution trials, following not only the total phenolic compounds but also certain individual ingredients. It was found that during loading from apple juice, the more hydrophobic ingredients were adsorbed to a larger extent than the more hydrophilic ingredients. During elution, the more hydrophilic ingredients were eluted to a larger extent than the more hydrophobic ingredients when water or 25:75 methanol-water mixtures were used as eluents. All these ingredients were eluted more completely however at high proportions of methanol in the methanol-water mixture eluents. In the long term, these results may guide to operating conditions that facilitate the separation of the phenolic compounds and the recovery of certain ingredients in a pure condition by selective adsorption and elution.

Anthocyanins can be recovered from grape pomace using synthetic adsorbents. These polyphenols are first extracted with acidified alcohols or with bisulfite solution in 40:60 ethanol-water to extract anthocyanins-HSO_3 complex. Recently, extraction is done without sulfitation but by enzyme assisted extraction to avoid the use of sulfites. After filtration, the solution passes through a DVB type adsorbent which adsorbs the anthocyanins and other phenolic compounds, typically at a specific flow rate 2-5 BV/h. Elution is achieved with methanol or ethanol acidified with HCl. Care should be taken for the swelling of the adsorbent in these solvents which varies depending on the adsorbent. In order to ensure a more concentrated eluate, three-column merry-go-round system can be used, two on loading and one on elution. The head column goes to elution when the second one breaks through. Anthocyanins are finally recovered from the concentrated eluate by distillation (Langston, 1985; Kammerer *et al*, 2005).

Recovery of phenolic compounds from grape and wine byproducts has been studied (Soto et al, 2012). Phenolic compounds were recovered from the liquid phase of distilled pomace. They were first concentrated by NF, followed by adsorption on polymeric adsorbents. Desorption was done with 96% ethanol at 50°C. A large number of commercial non-functionalized adsorbents varying in chemical nature, particle size, porosity, pore size and specific surface area, were evaluated for adsorption and desorption of phenolic compounds.

The adsorption and desorption of polyphenolic compounds of functionalized adsorbents were studied and compared to non-functionalized adsorbents (Silva et al, 2018). Styrene-DVB adsorbents were functionalized with imidazole or pyridine. The phenolic compounds used in the study were p-coumaeic acid, trans resveratrol and naringenin. The functionalized adsorbents shows significantly higher loading capaciries compared to non-functionalized. Formation of H-bonds formed with the nitrogen atoms of the imidazole or pyridine functional groups. Despite the higher adsorption, the elution with acidified alcoholic solution was easier for the functionalized adsorbents, possibly because at acidic conditions, the nitrogen atoms of the functional groups do not participate any more with H-bonds.

268

13.Miscellaneous applications in food processing

Synthetic sweeteners

Sucralose is a synthetic sweetenermade from sucrose by selective chlorination of the 4, 1' and 6' positions with a chlorinating agent (figure 13.1).

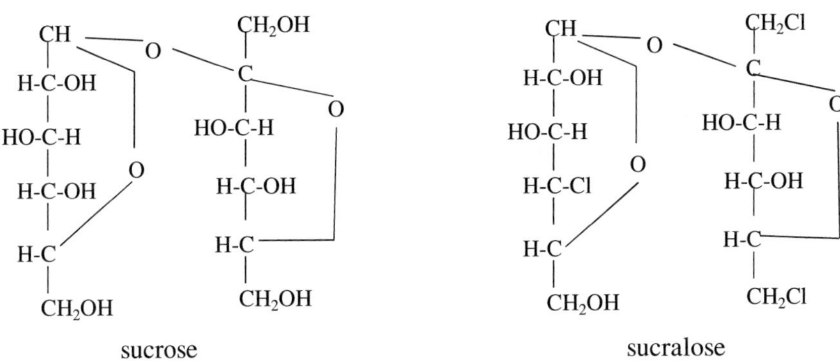

Figure 13.1 Chemical structures of sucrose and sucralose

This chlorination is achieved by protecting the primary hydroxyl groups of sucrose by acylation to form sucrose-6-acylate, chlorinating with a chlorinating agent to chlorinate the 4, 1' and 6' positions and deprotection of the primary alcohol groups to obtain sucralose (Wiley Jr *et al*, 2009). After chlorination, the intermediate product is quenched with NaOH giving NaCl salt. The concentration is high so that part of the salt precipitates out. The acylated sucralose is then deacylated with NaOH, neutralized with HCl, concentrated and the product is filtered to remove precipitated salts. The reaction medium is dimethylformamide (DMF) which during the manufacturing process decomposes partially to give dimethylamine (DMA).The remaining DMF is removed by steam stripping which leaves the remaining solution as an aqueous solution.Theaqueous solution is then diluted to bring down viscosity andtreated on a SAC exchange resin to recover DMA in order to convert it to DMF and recycle it back to the process.

The SAC resin operates in the Na^+ form. Since in the remaining solution there is high NaCl concentration, the DMA removal competes with the Na^+ present in solution, therefore the precipitate of the NaCl formed during the production of sucralose should be removed by filtration as much as possible. The resin is regenerated with NaCl or with NaOH solution to recover DMA and recycle it back to the process.

Pectins decalcification

Pectins are essentially partially methylated esters of polygalacturonic acid. The free carboxylic acid groups tend to expand their structure due to the repulsion of the negatively charged carboxylate groups (pKa of the carboxylate groups is about 2.9 there is therefore sufficient ionized groups). Divalent cations, like Ca^{2+}, can form bridges between two carboxylate groups situated in different polymer chains and can cause gelling. Pectins are soluble in water but insoluble in alcohols.

Methylation of the carboxylic acid groups forms esters that are more hydrophobic compared to free carboxylic acid and change the properties of the pectins. In fact, the properties of the pectins depend on the degree of esterification. Low methoxyl-pectins gel with Ca^{2+} bridging while high methoxyl- pectins do not, due to the lack of carboxylate groups. High methoxyl pectins gel by hydrogen bond formation and hydrophobic interactions in the presence of acids which suppress the dissociation of carboxylate groups.

Pectins are obtained from citrus peels and apple pomace by extraction with nitric acid. The acidic solution contains about 0.5-1% pectin, 2% acid and about 300-400 ppm of Ca^{2+} which needs to be removed. The Ca^{2+} removal is achieved with a SAC resin in the Na^+ form. Specific flow rate is about 4-6 BV/h, temperature 40-70°C. Operating capacity about 1 eq/L_R. Regeneration of the resin with 200 g NaCl/L_R using 10% NaCl solution.

Regeneration steps:
8) sweetening off, 2 BV soft water
9) air scouring
10) backwash, 1 hour or until effluents are clear.
11) settling
12) 10% NaCl regenerant, 2 BV, in 1 h.
13) Slow rinse, 2 BV/h, 1 h
14) Fast rinse, service flow rate, 0.5 h.

Wine stabilization.

While aging improve certain properties of wine, like aroma or taste, it may also degrade other properties such as clarity. In fact, the tartaric acid contained in the grapes forms potassium hydrogen tartrate (KHT) which during fermentation and the formation of alcohol, precipitates out with time. There exist a number of ways for preventing KHT from precipitating in the bottled wine. A common stabilizing method is chilling the wine at a temperature just above the freezing point over a period of two to three weeks. Chilling the wine decreases the solubility of KHT which crystallizes and precipitates out where it is removed by filtration before raising the temperature.

Ion exchange can be used to stabilize the wine from KHT precipitation. A SAC resin in the Na^+ form exchanges Na^+ for K^+ forming sodium hydrogen tartrate which is more soluble than KHT. However in this way, the level of Na^+ in the wine can increase to unacceptable levels. An alternative is to use the SAC resin in the H^+ form. Again, the resin removes all cations with an equivalent increase in acidity in the wine. The solution to this problem is to treat a fraction only of the wine and blend treated and untreated parts (Mira *et al*, 2006). Another way to minimize the decrease in the pH after the SAC resin in the H^+ form is to use astrong or weak base anion exchange resin following the SAC resin.

Typical specific flow rate is 20 BV/h. Regeneration is achieved with HCl used as 6-8% solution at a level of 80 g HCl/L_R or 4% NaOH at 70 g $NaOH/L_R$ in case WBA resin is used. If a SAC resin is used in the Na^+ form, regeneration is done with 10% NaCl solution at a level of 150 g $NaCl/L_R$.

Ion exchange for wine stabilization is authorized only in some countries specifying the types of resins, the ionic form and the specifications for the treated wines.

REFERENCES

Aehle W (Ed.) (2007). Industrial Enzymes *In* Enzymes in Industry Production and Applications, Third Edition, Wiley-VCH Verlag GmbH & Co. KGaA, Weinheim, Germany, p. 207-208.

Agalias A, Magiatis P, Skaltsounis AL, Mikros E, Tsarbopoulos A, Gikas E, Spanos I, Manios T (2007). A new process for the management of olive oil mill waste water and recovery of natural antioxidants. *J Agric Food Chem,* **55**: 2671-2676.

Albright LR, Jakovac IJ (1985). Catalysis by functionalized porous organic polymers. *IE 287-6* Rohm and Haas Company, Philadelphia, Pa.

Al Eid SM (2006). Chromatographic separation of fructose from date syrup. *Int J of Food Sci and Nutrition***57**: 83-96.

Alexandratos SD, Brown GM, Bonnesen PV, Moyer BA (2000). Bifunctional anion-exchange resins with improved selectivity and exchange kinetics. *US Patent 6,059,975 A.*

Arden TV (1968). *Water Purification by Ion Exchange.* Butterworth, London.

Arumugam BK, Collins NA, Macias TL, Perri ST, Powell JEG, Sink CW, Cushman MR (2003). Continuous process for producing L-ascorbic acid. *US Patent 6,610,863*

Asadi M (2007). *Beet-Sugar Handbook*, Wiley-Interscience.

Asher DR, Simpson DW (1956). Glycerol purification by ion exclusion. *J Phys Chem,* **60**: 518-521.

Ashurst PR (2002). Dowex Ion Exchange Resins Juice Encancement by Ion Exchange and Adsorbent Technologies. *The Dow Chemical Company*

Beadle JR, Saunders JP, Wadja TJ (1992). Process for manufacturing Tagatose. *US Patent 5,078,796.*

Bellamy S, Pease S (2010). Stabilization of liquid food and beverages. *EP Application EP 2 166 080*

Bennett AN (1945).recovery of betaine and betaine salts from sugar beet wastes. *US Patent 2,375,164.*

Bento LSM (1999). Process for regeneration of ion-exchange resins used for sugar decolorization, using chloride salts in a sucrose solution alkalinized with calcium hydroxide. *US Patent 5932106*

Binder TP (2001). Simulated Moving Bed Chromatographic Purification of Aminoacids. *European patent application EP 1 106 602*

Birbingham Jr BM (1967). Process for preparing purified glycerine. US Patert 3,563,954A.

Bissessur J, Permaul K, Oldhav B (2001).Reduction of patulin during apple juice clarification. *J Food Protection,* **64**: 1216-1219

Blaney LM, Cinar S, SenGupta AK (2007). Hybrid anion exchanger for trace phosphate removal from water and wastewater. *Water Res* **41***: 1603-13.*

Bonrath W (1999). Method of making d,l-alpha-tocopherol. *US Patent 5,900,494*

Bonrath W, Eggersdorfer M, Netscher T (2007). Catalysis in the industrial preparation of vitamins and nutraceuticals. *Catalysis Today,* **121**: 45-57.

Bouchard EF, Friedman IJ, Taylor RJ (1958). Vitamin B12 products and preparation thereof. *US Patent 2,830,933.*

Burris B, Bathany Y, Paananen H (2009). Improved betaine recovery during molasses desugarization using the NS2Pchromatography process. *Proceedings from the 35th meeting of the American Society of the Sugar Beet Technologists, Orlando, Fla,*

Burkhardt MO, Schick R, Freudenberg T (2000). Continuous thin juice decalcification with weakly acidic cation exchangers in Pfeifer & Langen's Appeldorn factory.*Zuckerindustrie,***125**: 673-682.

Byung YY, Rex M (1996). Alkaline degradation of glucose: effect of initial concentration of reactants. *Carbohydrate Res,* **280**: 27-45 and 47-57.

Chung Y, Chu O, Alvarez MP (2007). Resin deacidification of citrus juice and high acid maintenance. *US Patent 7264837*

Clarke RJ, Vitzthum OG (2008). *Coffee: Recent Developments,* John Wiley and Sons, Ch. 5 p. 117

Clifford DA (1999). Ion Exchange and Inorganic Adsorption. *In* Letterman R *Water Quality and Treatment*. McGraw Hill, Inc., New York, 5th Edition, 9.1-9.91.

Coca M, Mato S, Gonzales-Benito G, Uruena A, Garcia-Coubero T (2010), Use of weak cation exchange resin Lewatit S 8528 as alternative to strong ion exchange resins for calcium salt removal.*J Food Eng,***97** : 569-573.

Celle R, Hervé D (1980). Décoloration du sirop de refonte du sucre de canne à la raffinerie de Marseille. *Ind Alimentaires Agricoles,* Juillet-Août N° 7-8

Chassagne P, Khanzhin N, Matwiejuk M, Hederos M, Banke N (2017). Separation of oligosacchrides from fermentation broth. *European Patent WO 2017/182965 A1.*

Chen JCP (1985). *Cane Sugar Handbook*. John Wiley & Sons, Eleventh Edition.

Cousin MA, Riley RT, Pestka JJ (2005). Foodborne pathogens: Chemistry, Biology, Ecology and Toxicology. *In*: Fratamico PM, Bhunia AK and Smith JL *Foodborne pathogens: microbiology and molecular biology*, Caister Academic Press

Crawford TC, Crawford SA (1980). Synthesis of L-ascorbic acid. *In:* Tipson RS, Horton D *Advances in Carbohydrate Chemistry and Biochemistry*. Academic Press Inc, 79-155.

Das K, Anis M, Azemi BMN, Ismail N (1995). Fermentation and recovery of glutamic acid from palm waste hydrolysate by Ion-exchange resin column. *Biotech and Bioeng,***48**: 551-555

Datta R, Glassner DA, Jain MK, Vick Roy JK (1992). Fermentation and purification process for succinic acid. *US Patent 5,168,055.*

Davankov VA, Tsyurupa MP (1989). Structure and properties of porous hypercrosslinked polystyrene sorbents 'Styrosorb'. *Pure Appl Chem,* **61**: 1881-1888.

de Dardel F, Arden TV (1989). Ion Exchangers. In *Encyclopaedia of Technical Chemistry.*VCH, Weinheim, Germany Vol. A14

de Dardel F (2007). Les résines échangeuses d'ions en traitement d'eau potable. *L'eau, L'industrie, Les nuisanses N°306, 59-64.*

Deleyn F, Aiken C, Derez F, Mauro D, Provoost D, Stalin M-O, Vanhemelrijck B (2012). Process for the production of maltodextrins. *European Patent EP 1811863 A1.*

De Wit JN (2001). Lecturer's handbook on whey and whey products.*European Whey Products Association, Brussels, Belgium*

Diaion® Ion Exchange Resins Manual, 1995, Vol. 1, p. 92, Mitsubishi Chemical Corporation

Di Mauro A, Arena E, Fallico B, Passerini A, Maccarone E (2002). Recovery of anthocyanins from pulp wash of pigmented oranges by concentration on resins. *J Agric Food Chem,***50**: 5668-5674

Dultani S, Turhan KN, Etzel MR (2004). Fractionation of proteins from whey by using cation exchange chromatography. *Process Biochem.* **39***: 1737-1743*

Ellis VA, Wilson MA (2002). Carbon Exchange in Hot Alkaline Degradation of Glucose. *J Org Chem* **67** *(24): 8469-8474.*

Fechter WL, Dienst JH, Le Patourel JF (1997). Recovery of aminoacid. US Patent 5,684,190

Fechter WL, Kitching SM, Reimann RH, Ahmed FE, Jensen CRC, Schorn PM, Walthew DC (2001). Direct production of white sugar and whitestrap molasses by applying membrane and ion exchange technology in a cane sugar mill. *24th ISSCT Congress*, 100-107.

Felber. EH (1971), New process for deliming thin juice without regeneratingagents and waste water. *Amer Soc Sugar Beet Technol* **16**:279-288.

Felton GE (1949). Ion Exchange Application by the Food Industry. *In:* Mrak EM, Stewart GF *Advances in food research Vol. 2.* Academic Press

Fernandez-Bolanos J, Rodriguez G, Gomez E, Guillen R, Jimenez A, Heredia A, Rodriguez R, 2004. Total recovery of the waste of two-phase olive oil processing: isolation of value added compounds. *J Agric Food Chem,***52**: 5849-5855.

Fernandez-Bolanos J, Moreno AH, Rodriguez G, Rodriguez R, Jimenez A, Guillen R, 2005. Method for obtaining purified hydroxytyrosol from products and by-products derived from the olive tree. *US Patent 6,849,770.*

Fernandez-Bolanos J, Rodriguez G, Rodriguez R, Guillen R, Jimenez A (2006). Extraction of interesting organic compounds from olive oil waste. *Grasas y aceites,***57**: 95-106.

Franck A, De Leenheer L (2005). Inuline. *In*: Steinbüchel A, Rhee SK *Polysaccharides and polyamides in Food Industry Vol. 1*. Wiley-VCH.

Gattuso G, Barreca D, Gargiulli C, Leuzzi U, Caristi C (2007). Flavonoid composition of citrus juices. *Molecules,***12**: 1641-1673.

Gerberding SJ, Singh R (2012). Purification of succinic acid from the fermentation broth containing ammonium succinate. *European Patent EP 2519491 A2*.

Ghelmetti G, Trinchera C (1970). Process for the production of Lysozyme. *US Patent 3,515,643*.

Giovanetto RH (1990). Method for the recovery of steviosides from plant material. *US Patent 4,892,938*.

Guidini CZ, Fischer J, Soares Santana LN, Cardoso VL, Ribeiro EJ (2010). Immobilization of Aspergillus orizae β-galactosidase in ion exchange resins by combining ionic binding method and crosslinking. *Biochem Eng J* **52**: 137-142.

Gluszcz P, Jamroz T, Sencio B, Ledakowicz S (2004). Equilibrium and dynamic investigations of organic acids adsorption onto ion exchange resins. *Bioprocess Biosyst Eng,***26**: 185-190.

Gogoi P, Dutta NN, Rao PG (2010). Adsorption of catechin from aqueous solutions on polymeric resins and activated carbon. *Indian J Chem Technol,* **17**: 337-345.

Gomes FNDC, Pereira LR, Ribeiro NFP, Souza MMVM (2015). Production of 5-hydroxymethylfurfural (HMF) via fructose dehydration:effect of solvent and salting-out. *Braz J Chem Eng* **32: 501-508**.

Gregor HP, Hamilton MJ, Oza RJ, Bernstein F (1956). Studies on ion exchange resins. XV. Selectivity coefficients of methacrylic acid resins toard alkali metal cations. *J Phys Chem* **60:** 263-267.

Groom DR, Jarski H, McGillivray TD (2009). Optimation of betaine recovery in a coupled loop molasses desugarization separator. *Proceedings from the 35th meeting of the American Society of the Sugar Beet Technologists, Orlando, Fla,*

Gryllus E, Delavier HJ (1975). Das BMA-Zsigmand-Gryllus-Verfahren ein neues Verfahren zur Ohnsaftentkaltung Teil 1: Prinzipielles der die Ohnsaftenkaltung, *Zukerindustrie* **25**: 493-500.

Gryllus E,Eszterle M,Anyos E, Daminati M (1998).The RDN regeneration method for decalcification of thin juice. *Zuckerindustrie,* **123**: 36-43.

Gu B, Brown GM, Maya L, Lance MJ, Moyer BA (2001). Regeneration of perchlorate (ClO_4^-)-loaded anion exchange resins by a novel tetrachloroferrate ($FeCl_4^-$) displacement technique. *Environ Sci Technol **35: 3363-3368.***

Gu B, Brown GM (2006). Recent Advances in Ion Exchange for Perchlorate Treatment, Recovery and Destruction. *In* Gu B and Coates JP, Editors, *Perchlorates, Environmental Occurrence, Interactions and Treatment. Springer 2006, p. 209-252.*

Hajek M, Skopal F (2010). Treatment of glycerol phase formed by biodiesel production. *Bioresources Tech,***101**: 3242-3245

Harju M (2007). Chromatographic separation of lactose and its applications in the dairy industry. *IDF Symposium Lactose and its derivatives, Moscow, 14-16.5.2007*

Heding LG and Gupta JK (1975). Improvement of conditions for precipitation of citric acid from fermentation mash. *Biotechnology and Bioengineering,***17**: 1363-1364.

Heikkila HO, Melaja JA, Millner DED, Virtanen JJ (1985). Betaine recovery process. *EP0 054 544B1*

Helfferich F (1961). "Ligand exchange": A Novel Separation Technique. *Nature **189** : 1001-1002.*

Helfferich F (1962). *Ion Exchange.* McGraw-Hill Book Company, Inc.

Helfferich F (1966) Ion-exchange kinetics. *In* Marinski J *Ion exchange, a series of advances*, 65-100

Henscheid TH, Velasquez L, Meacham D (1990). Weak cation softening of thin juice. *International Sugar J*, **92**: 206-209.

Hitoh H, Toide K, Ikeda M (1994). Method for purification of an aminoacid using ion exchange resin. *US Patent 5279744*

Jennewein S (2015). Process for efficient purification of neutral human milk oligosaccharides (HMOs) from microbial fermentation. *European Patent EP2896628 A1*

Jonnala KK, Kiran BGD, Kaul VK, Ahuja PS (2006). Process for production of steviosides from Stevia Rebaudiana Bertoni. *US Patent Application 2006/0142555 A1.*

Kammerer DR, Saleh ZS, Carle R, Stanley RA (2007). Adsorptive recovery of phenolic compounds from apple juice. *Eur Food Res Technol,* **224**: 605-613.

Kammerer DR, Gajdos Kljusuric J, Carle R, Schieber A (2005). Recovery of anthocyanins from grape pomace extracts(Vitis vinifera L. Cv. Cabernet Mitos) using a polymeric adsorber resin. *Eur Food Res Technol,* **220**: 431-437.

Kammerer J, Carle R, Kammerer DR (2011). Adsorption and ion exchange: basic principles and their application in food processing. *J Agric. Food Chem.* **59**: 22-42.

Karrenbauer M, Kleemann A, Leuchtenberger W, Moerck R (1988). Process for the recovery of malic acid. *US Patent 4,772,749.*

Kawai K, Makino K, Tamura M and Tanimura M (1998). Process for producing purified L-ascorbic acid. *US Patent 5,817,238*

Klipper R, Wagner R, Soest KH and Litzinger U (2007). Polyol refining.*Patent application 12027355,2008-02-07*

Kochergin V, Kearney M, Jacob W, Velasquez L, Alvarez J, Baez-Smith C (2000). Chromatographic desugarization of syrups in cane mills.*Indian Sugar* **50** : 585-597.

Kola O, Kaya C, Duran H and Altan A (2010). Removal of Limonin Bitterness by Treatment of Ion Exchange and Adsorbent Resins. *Food Sci. Biotechnol.* **19**: 411-416

Kuellmer V (2001). Ascorbic Acid. *Kirk-Othmer Encyclopedia of Chemical Technology.* John Wiley and Sons, Inc

Kulprathipanja S, Oroskar AR (1991). Separation of an organic acid from a fermentation broth with an anionic polymeric adsorbent. *US Patent 5,068,419.*

Kulprathipanja S(1986). Separation of citric acid from fermentation broth with a neutral polymeric adsorbent. *US Patent 4,720,579.*

Kun KA, Kunin R (1967). The pore structure of macroreticular ion exchange resins. *J Polymer Sci Part C: Polymer Symposia,* **16** : 1457-1469.

Kunin R (1958). *Ion Exchange Resins.* John Wiley, 2nd Edition.

Kwok RJ Pacific: UF/Softening of clarified juice: the door to direct refining and molasses desugarization in the cane sugar industry. Proc S African Sugar Technologists Association N° 70 p. 166-170.

Laksameethanasan P, Somla N, Janprem S, Phochuen N (2012). Clarification of sugarcane juice for syrup production. *Procedia Engineering* **32:** 141-147.

Lancrenon X, Hervé D (1988). Recent trends in the use of ion exchange in the sugar industry. *Sugar Technology Reviews,* **14:** 207-274.

Lancrenon X, Paillat D (2005). Production of liquid sugars from fine liquor in the refinery. *Sugar Industry Technologists Meeting, Dubai,* UAE, paper # 870.

Lancrenon X, and Fedders J (2008). An innovation in glycerine purification. *Biodiesel Magazine, June 2008*

Langston M (1985). Anthocyanine colorant from grape pomace. *US Patent 4,500,556.*

Lineback DS, Chu OA, Chung Y, Pepper MA, Alvarez MP (2006). Juice deacidification. *US Patent 7,074,448*

Laufenberg G, Kunz B, Nystroem M (2003). Transformation of vegetable waste into value added products: (A) upgrading concept (B) practical implementations. *Bioresource Techn,* **87**: 167-198.

Li-Chan E, Nakai S, Sim J, Bragg DB, Lo KV (1986) Lysozyme separation from egg white by cation exchange column chromatography. *J Food Sci 51: 1032-1036.*

Lyndon RM, Miller CJ (2001). Process for reducing the patulin concentration in fruit juices. *US Patent 6248382*

MacLaurin DJ, Green JW (1969). Carbohydrates in alkaline systems. I. Kinetics of the transformation and degradation of D-glucose, D-fructose, and D-mannose in 1 M sodium hydroxide a t 22 °C. *Canadian J of Chem,* **47**: 3947-3955.

Maestri DM, Nepote V, Lamarque AL, Zygadlo JA, 2006. Natural products as antioxidants. *In* F Imperato *Phytochemistry: Advances in Research 2006,* Research Signpost Publication, Ch. 5, 105-135.

Margolis G, Rushmore DF, Liu RT-S (1977). Decaffeination uf vegetable material. *US Patent 4,03,251*

Martin BD, Parsons SA, Jefferson B (2009). Removal and recovery of phosphate from municipal wastewaters using a polymeric anion exchanger bound with hydrated ferric oxide nanoparticles. *Water Sci Technol 60: 2637-45.*

Mavrakis T (2009). *Exploitation of bioactive constituents of olive leaves, grape pomace, olive mill waste water and their applications in phytoprotection*. PhD Thesis, Cranfield University.

Melaja AJ, Hamalainen L (1975). Process for the production of mannitol and sorbitol . *US Patent 3864406*

Milnes B and Agmon G (1995). Debittering and Upgrading Citrus Juice and By-products using Combined Technology. Presented at the citrus processing short course, September 1995 (sponsored by the Institute of Food Technologists –Florida Section) in Clearwater, FL.

Mindler AB (1949). Sugar Refining and By-product Recovery. *In* FC Nachod *Ion Exchange Theory and Application*, Academic Press Inc., 315-350.

Mira H, Leite P, Ricardo-da-Silva JM, Curvelo-Garcia AS (2006). Use of ion excha,ge resins for tartrate wine stabilization. *J Int Sci Vigne Vin***40** : 223-246.

Moroe T, Hattori S and M, Komatsu A, Matsui T, Kurihara H (1969). Process for producing α-tocopherol and its esters. *US Patent 3,459,773*

Mottard PL (1983). The Imacti process for juice decalcification. *Int Sugar J***85**: 233-237.

Nasab EE, Habibi-Rezaei M, Kakhi A, Balvardi M (2009). Investigation on acid hydrolysis of inulin: a response surface methodology approach. *International J of Food Eng***5**, iss.3, art. 12

Nelan DR, Foster CH (1980). Conversion of non-alpha-tocopherols to alpha-tocopherols. *US Patent 4,239,691.*

Norman SI (1999). Juice enhancement by ion exchange and adsorbent technologies. *In* Ashurst PR*Production and packaging of Non-Carbonated fruit juices and Fruit Beverages,* Aspen Publication, 253-272.

Novotny O, Cejpek K, Velisek J (2008). Formation of carboxylic acids during degradation of monosaccharides. *Czeck J of Food Sci 26: 117-131.*

Olivério LJ, Boscariol F (2006). DRD – Dedini Refinado Direto (Dedini Direct Refined) : The refined sugar without a refinery. *Intern. Sugar J***108**: 239-246

Paananen H, Saari P, Nurmi N (2007). Method for separating betaine. *US Patent Application 2007/0158269 A1*

Pannekeet W (1982). Process for the regeneration of a sorbent. *US Patent* 4,353,992

Pannekeet W (1984). Novel Regeneration Systems for Ion-Exchange Resins in the Sugar Industry.*Solvent Extraction and Ion Exchange,* **2**: 741-754

Parajo JC, Dominguez H, Moure A, Diaz-Reinoso B, Conde E, Soto ML, Conde MJ, Gonzales-Lopez N, 2008. Recovery of phenolic antioxidants released during hydrolytic treatments of agricultural and forest residues. *EJEAFChe***7**: 3243-3249.

Payzant JD, Laidler JK, Ippolito RM (1999). Method of extracting selected sweet glucosides from the stevia rebaudiana plant.*US Patent 5,962,678.*

Pazouki M, Panda T (1998). Recovery of citric acid-a review. *Bioprocess Biosyst. Eng.* **19**: 435-439

Pitochelli AR (1975). Ion exchange catalysis and matrix effects. Rohm and Haas Company, Philadelphia, Pa.

Prielipp GE, Keller HW (1955). Purification of crude glycerine by ion exclusion. *J American Oil Chemists' Soc.* **33**: 103-108

Ramchander S, Feather MS (1974). Studies on the mechanism of color formation in glucose syrups. *Cerial Chem* **52:** 166-172.

Reckziegel A, Rezkallah A, Mir F (2018). Strong base polyacrylate anion exchangers. *European Patent Application 3 308 857 A1*.

Reed Jr SF (1981). Polymeric adsorbents from macroreticular polymer beads. *US Patent 4263407*

Rein PW, Bento LSM, Cortes R (2007). The direct production of white sugar in a cane sugar mill. *Intern SugarJ***109**: 286-299

Ribeiro RR, Vitolo M (2005). Anion exchange resin as support for invertase immobilization. *J Basic and Appl Pharmaceutical Sci***26:** 175-179.

Riegel M, Höll WW (2008). Removal of natural uranium from groundwater by means of weakly basic anion exchangers. *In* Cox M, *Recent Advances in Ion Exchange Theory and Practice.* (proceedings of IEX2008). SCI, 2008, 331-338.

Romano P (2018). Experiences in hexavalent chromium removal in the treatment of drinking water. *In* A.Gilardoni Ed. *The Italian Water Industry: Cases of Excellence.* Springer International Publishing AG, 101-107

Rossiter G, Jensen C, Fechter W (2002). White sugar at the cane factory-Impact of WSM on factory operations. *SPRI Conference.*

Rousseau G (1999), Method of regenerating ion exchange resins in the process of decalcification of sugar factory juices. *EP 5,958,142*

Rousseau G (2003), Process for regenerating ion exchange resins in the process of decalcifying sugar juices.*EP 0832986*

Rousseau G, Rovel B and Metche M (1990). Cristallisation et qualité des sucres. *Ind Alimentaires et Agricoles***107** : 629-636.

Saari P, Häkkä K, Jumppanen J, Heikkilä H, Hurme M (2010).Study on Industrial Scale Chromatographic Separation Methods of Galactose from Biomass Hydrolysates. *Chem Eng and Technology* **33**: 137-144

Salome JP, Ferez P, Lefevre P (2002). Process for preparing a non-crystallizable polyol syrup. *US Patent 6,417,346.*

Sarkar S, Smith RC, Blute N, SenGupta AK (2016). Removal of trace chromate from contaminated water: ion exchange and redox-active sorption processes. *In* Bryjak M, Kabay N, Rivas BL, Bundschuh J, Editors, *Innovative Materials and Methods for Water Treatment: Solutions for Arsenic and Chromium Removal.* CRC Press, 2016, p.297-326.

Scordino M, Di Mauro A, Passerini A, Maccarone E (2003). Adsorption of flavonoids on resins: Hesperidin. *J Agric Food Chem,***51**: 6998-7004.

Scordino M, Di Mauro A, Passerini A, Maccarone E (2004). Adsorption of flavonoids on resins: cyanidine-3-glucoside. *J Agric Food Chem,***52**: 1965-1972.

Scordino M, Di Mauro A, Passerini A, Maccarone E (2005). Selective recovery of anthocyanins and hydroxycinnamates from by-product of citrus processing. *J Agric Food Chem,***53**: 651-658.

Scordino M, Di Mauro A, Passerini A, Maccarone E (2007). Highly purified sugar concentrate from a residue of citrus pigments recovery process. *LWT-Food Sci Technol,***40**: 713-721.

Shaw PE, Tatum JH, Berry RE (1967). Acid-catalyzed degradation of D-fructose. *Carbohydrate Research* **5:** 266-273.

Shaw PE, Baines L, Milnes BA, Agmon G (2000). Commercial Debittering Processes to Upgrade Quality of Citrus Juice Products. *In* Berhow MA, Hasegawa S, Manners GD, *Citrus Limonoids Functional Chemicals in Agriculture and Food,* ACS Symposium series Vol. 758, Ch. 9, 120-131.

Siebert KJ (1999). Effects of Protein–Polyphenol Interactions on Beverage Haze, Stabilization and Analysis*J Agric Food Chem*, **47**, 353–362.

Silhankova L (1977). Method for preparation of Vitamin B1. *US Patent 4,186,252A.*

Silva M, Castellanos L, Ottens M (2018). Capture and purification of polyphenols using functionalized hydrophobic resins. Ind *Eng Chem Res 57(15): 5359-5369.*

Soto ML, Conde E, Gonzáles-Lopéz N, Conde MJ, Moure A, Sineiro J, Falqué E, Domíngez H, Núñez MJ, Parajó JC (2012). Recovery and concentration of antioxidants from winery wastes. *Molecules 17(3): 3008-3024.*

Sumikawa S, Maida R, Kageyama Y (1976). Process for the purification of malic acid. *US Patent 3,983,170.*

Tadashi A and Eiji I (2004). Fundamental characteristics of synthetic adsorbents intended for industrial chromatographic separations. *J Chromatography A,* **1036**: 33-44.

Théoleyre MA and Baudouin S (2007). Method for purification of an organic acid by chromatography. *WIPO Patent application*WO/2007/128918.

Théoleyre MA and Gula F (2004). Purification of food streams by combining ion exchange and membranes technologies: Application of decalcification in the whey industry. *International Conference Engineering and Food, ICEF9- 2004.*

Tomotani EJ, Vitolo M (2006). Catalytic performance of invertase immobilized by adsorption on anionic exchange resin. *Process Biochem***41**: 1325-1331.

Top AGM, Leong WL, Ong ASH, Kawada T, Watanabe H, Tsuchiya N (1993). Production of high concentration tocopherols and tocotrienols from palm oil byproducts. *US Patent 5,190,618.*

Wang JC, Cui F, Li T (2012) Optimization of Synthesis Process for Sodium Ascorbate. *Advanced Materials Research **550-553:** 10-15.*

Van Lancker F (2007). Process for preparing alkali- and heat-stable sugar alcohol compositions and a sorbitol composition. *US Patent 7,179,336.*

Van Walsem HJ, Thompson MC (1997). Simulated moving bed in the production of lysine. *J Biotech* **59** : 127-132.

Vera E, Dorrnier M, Ruales J, Vaillant F, Reynes M (2003). Comparison between different ion exchange resins for the deacidification of passion fruit juice. *J. Food Eng,***57**: 199-207.

Wiley Jr JE, Leinhos DA, Dentel DA, Kerr J (2009). Method for the production of sucralose. *US Patent Application 20090247737.*

Wolfgang J, Prior A, Bart HJ, Messenböck RC, Byers CH (1997). Continuous separation of carbohydrates by ion exchange chromatography. *Separation Sci and techn,***32**: 71-82.

Wu J, Peng Q, Arlt W and Minceva M (2009). Model-based design of a pilot-scale simulated moving bed for purification of citric acid from fermentation broth. *J of Chromatography A,***1216**: 8793-8805

Wu SH, Yu W, Liao L, Wei S and Zeng L (2004). Study of separating and abstracting L-Leucine from fermentation liquor. *In* Tong Z, Kim SH*Proceedings of the 4th international conference on separation science and technology,*786-792.

Yi YB, Lee JW, Hong SS, Choi YH, Chung CH (2011).Acid-mediated production of hydroxymethylfurfural from raw plant biomass with high inulin in an ionic liquid. *J Ind Eng Chem***17**: 6-9.

Yu J, Duan Z-K, Yang Y-Q, Tang X (2006). Purification of natural Vitamin E by adsorption. *China Oils and Fats 2006-10.*

Zaganiaris EJ (2017). *Ion exchange resins in uranium hydrometallurgy, 2nd edition.* Books on Demand.

Subject Index

A

B

C

F

Fatty acids   243-245, 247, 259
Flavanols   260, 261
Flavonoids   167, 188, 260, 264
Flavonols   260, 261
Fluorides (removal from drinking water)   210
Free water   25
Freundlich   54, 266, 267
Fructose   157-161, 164-166, 169
Fumaric acid   236-237

G

Galactose   176-177
Gelatin   239-242
Glucose demineralization   147-158
Glucose-fructose separation   158-161
Glucose isomerase   61, 158
Glucose isomerization   157-158
Glutamic acid   223-224
Glycerin   243-247
Grape juice   190-191
Gryllus process   108-109

H

Haze
     apple juice   182-183
     beer   193
Hesperidine   188, 264, 265, 266
Higgins Loop®   82-83

L

M

N

Z